Water 4.0

The book guides readers through the key issues within water supply systems, and how they have seen several technological advancements over the years. It examines the application of the Fourth Industrial Revolution to water systems operation with the advent of Big Data and computational information. The importance of interconnectivity and data analytics through automation, digitalisation, and machine learning is also explored for scaling up water supply operations.

Features:

- Examines the application of the Fourth Industrial Revolution to water systems operation.
- Applies the concept of intelligent computing through the exploration of Big Data and modelling to water supply network management.
- Discusses the application of automation in water supply network management with practical examples.
- Explores the use of automatic and electronic devices to monitor and control water loss.
- Reviews the historical development of technology from the First to the Fourth Industrial Revolution.

The book is aimed at graduate and postgraduate students, and researchers in water and environmental engineering, policymakers, and industry practitioners.

Water 4.0

Concepts and Application

Adeoba Mariam Iyabo and Nnamdi Nwulu

CRC Press
Taylor & Francis Group
Boca Raton London New York

CRC Press is an imprint of the
Taylor & Francis Group, an **informa** business

First edition published 2027
by CRC Press
2385 NW Executive Center Drive, Suite 320, Boca Raton FL 33431

and by CRC Press
4 Park Square, Milton Park, Abingdon, Oxon, OX14 4RN

CRC Press is an imprint of Taylor & Francis Group, LLC

ISBN: 9781041031567 (hbk)
ISBN: 9781041061120 (pbk)
ISBN: 9781003633839 (ebk)

DOI: 10.1201/9781003633839

Typeset in Times
by codeMantra

Contents

Preface

Water remains an essential need for nature, including humanity, to survive. Despite its abundance, the mode and means of transportation and transmission to users have always been problematic. The need for various water supply systems improvement and development that facilitate reliability and efficiency, fit for all purposes, is continuous. The outburst of the world's population and drastic climatic change will further pressure the necessity of more efficient and rapid water supply systems.

This book takes the reader through the key issues with water supply systems and how several technological changes have been seen over the years and generations. It further exposes the readers to some of the water network management application techniques available for operations and the need to embrace the evolution of the Fourth Industrial Revolution (Industry 4.0). Some of the ongoing advancements in Industry 4.0 applicable to water (Water 4.0) are fully expanded for readers to understand.

This book will be helpful to graduate and postgraduate students, researchers, policymakers, and industry practitioners.

About the Authors

Adeoba Mariam Iyabo is an accomplished researcher and academic, holding a Ph.D. in Zoology from the University of Johannesburg and an M.Sc. and a B.Sc. (Honours) from the University of Agriculture, Abeokuta, Nigeria. She is currently a senior research associate at the University of Johannesburg's Centre for Cyber-Physical Food, Energy and Water Systems. In addition, she serves as a lecturer in Water Resources Management at the Pan African University for Life and Earth Science Institute, University of Ibadan, Nigeria, as well as a research fellow with the Biomechanics Research Group at the University of South Africa. Drawing on her multidisciplinary background, her research spans environmental management, freshwater biodiversity conservation, renewable energy systems, and sustainable water resources management, with a particular focus on integrating artificial intelligence into sustainability initiatives and exploring the interconnections within the Food-Energy-Water nexus. She contributes actively to the scientific community as an editor of the *Marine and Environment* journal and as a reviewer for multiple scholarly publications. She has also mentored numerous postgraduate students across different universities in Nigeria and South Africa. Beyond her academic roles, she co-founded the Conserve Africa Initiative, a nonprofit organisation dedicated to environmental preservation and community empowerment, reflecting her commitment to public engagement and sustainable development. Her work has earned recognition through prestigious honours, including a Canadian Science Publishing Award for one of the top ten papers of 2019, as well as research grants and merit awards from the University of Johannesburg, underscoring her impact in the field.

Nnamdi Nwulu is a Professor of Electrical Engineering and the Director of the Centre for Cyber-Physical Food, Energy and Water Systems at the University of Johannesburg. His research interests have focused on applying digital technologies, mathematical optimisation techniques, and machine learning algorithms in food, energy, and water systems. This research focus has produced five books and over 70 indexed journal publications, winning him the 2021 NRF Research Excellence Award for Early Career/Emerging Researchers in the Engineering/Technology Category. He was also a 2020/2021 TW Kambule-NSTF-South 32 Award finalist in the Emerging Researcher category and won the 2020 UJ Vice Chancellor's Most Promising Young Researcher Award. He is a member of the South African Young Academy of Science. He is a professional engineer registered with the Engineering Council of South Africa), a senior member of the Institute of Electrical and Electronics Engineers, a senior member of the South African Institute of Electrical Engineers, and a Y-rated researcher by the National Research Foundation of South Africa. He is the Editor-in-Chief of the *Journal of Digital Food Energy and Water Systems* and Associate Editor of the *IET Renewable Power Generation* and the *African Journal of Science, Technology, Innovation and Development.*

1 Introduction to the Water Supply System

1.1 INTRODUCTION

Water is one of the most important resources on Earth, which is why many cities and towns were more expansive near oceans and rivers; they realised the essentiality of the resource. The preservation of water supplies and the continuation of life on Earth should be primary concerns for humankind in the 21st century, and one of our primary goals should be to safeguard water quality (Pietrucha-Urbanik et al., 2020; Tzanakakis et al., 2020). The rising requirement for alternative water supplies to combat the current state of water scarcity, whether to aid food production and utilisation of localised water supply, industry utilisation, or consumption, highlights the necessity to put precautions into effect. Consideration must be given to advancements in technology in treating water and revising established recycling standards across various recycling categories to incorporate emerging contaminants (Tzanakakis et al., 2020). For emerging nations to enhance treatment and provide healthy potable water to their populations, it is crucial to construct and execute affordable hygiene and household-centred cleanliness systems, particularly in remote regions (Tzanakakis et al., 2020). Hence, it becomes important to establish innovative scientific tools and methods to meet the hurdles posed by rising populations and climate change.

A water supply system (WSS) encompasses the centralised, organised structure of distributing water. Water remains a universal asset that requires exceptional management and conservation amid ongoing crises, such as the increasing human population and rapid climate change. The surge in human population in this century calls for urgent measures to provide quality water fit for humankind and to ensure that it is adequately supplied without any form of wastage. According to Sustainable Development Goal (SDG) 6, clean and safe water for all is proposed for 2030. To attain this goal, many stakeholders' responsibilities need to be aligned. The participation of governments, private companies, non-governmental organisations, and the public at all levels is inherent. For most countries, the government plays a more significant role in the coordination of water supply by privatising units. According to Abebe and Tucho (2021), access to safe and clean water is dependent on and not limited to multiple factors, such as social class, creating a public consciousness, and level of exposure. Effective WSS management is not achieved in isolation but rather with an intertwined collaboration between the stakeholders in society.

Water usage and wastage have been attributed to individual and corporate lifestyles in different countries. Access to the water system varies from nation to state and is largely dependent on the socioeconomic status. Most developing countries do not have the luxury of good WSSs, due to prevailing systemic challenges varying from political to socioeconomic development, the level of awareness, sociocultural

DOI: 10.1201/9781003633839-1

factors, level of education, and other conditions. One of the resulting conditions of wasting water is water shortage. This book chapter will detail the domain in which water supply network management can be applied.

1.2 WATER RESOURCE, CLIMATE CHANGE, AND WATER SCARCITY

Globally, water remains one of the abundant compounds on and under the Earth's surface. Water resource availability depends on sustainable good water sources (Stringer et al., 2021). Many factors, such as increasing population, urbanisation, industrialisation, and climate change, are responsible for the scarcity of water resources. Climate change has become one of the significant drivers of water stress, and its ripple effects are seen in water scarcity and shortage. As climate changes rapidly and population grows steadily, water demand increases. Shortages in rainfall, precipitation, and higher temperatures are leading to significant pressure of climate change on water scarcity. The outcome of high water demand has triggered extreme water scarcity in many areas with surging populations.

According to UN Water Development (2020), there are proven track records of how climate change will continually affect water availability, quality, and quantity in many societies. Water shortage will affect all other primary SDG sectoral goals, such as health, food security, energy, and education. Water scarcity requires proactive measures from all stakeholders to solve it in a sustainable manner, such that the level of supply is well-balanced with the high demand. In many water-scarce environments, many strategies have been explored to conserve water resources with better infrastructures, planning, and management. These measures differ from one world region and country to another.

1.3 WATER SUPPLY SYSTEM OPERATION

Most of the available potable drinking water worldwide is derived from surface water. Other water sources, such as rain, spring, and groundwater are available. The common characteristics of all the water sources remain collection, storage, and transmission to a larger community for adequate usage. The primary aim of a WSS operation is to distribute good quality water without contamination to benefit livelihood and the environment. Over time, the WSS has evolved with several technologies to handle successful delivery and distribution. In maintaining and ensuring smooth operation, many systems need to work interdependently with the effectiveness and choice of human resource and utility managers (Bello et al., 2019). In achieving effective distribution, many inherent challenges are common in many countries.

1.4 KEY PROBLEMS IN WATER SUPPLY SYSTEM MANAGEMENT

The key stakeholders in many countries and regions depend variably on the municipal WSS for their citizens. In doing this, many WSSs depend on human management for hitch-free water availability. These management and operation challenges come in different forms and risk mitigating the plan for a constant water supply to communities. Some challenges are relatively managed with traditional management skills, and others are better solved with technological advancements (Bello et al., 2019). Some of such problems are further highlighted and discussed in the next section.

1.4.1 Water Loss Control

Water loss constitutes one major problem in any WSS. Water failures are associated with water loss through leaking pipes globally. Many factors can be accountable for leakages in the distribution system, which are also influenced by the severity and type of leakage involved. Leakages are categorised as small, medium, and high (Van Duuren et al., 1980). Water loss control primarily aims to conserve water and manage leakage to the minimal level possible within the water distribution networks (WDNs). In water loss management, it is imperative to possess a comprehensive understanding of the various components that play a role in mitigating issues within each distribution system. To effectively address water loss concerns cost-efficiently, it is imperative to possess the necessary skills, technological expertise, and operational and managerial methods. Besides leaking pipes and lack of knowledge, water misuse or excess usage, lowered tariff metering strategies, and political influences in many societies are responsible for water loss (Thornton et al., 2008). Real and apparent water losses can be controlled if all interdependence strategies are implemented appropriately. Otherwise, the unaccounted lost water will continually account for non-revenue water (NRW). The water audit approach, which has been jointly created by the International Water Association (IWA) and the American Water Works Association (AWWA), is advocated by the AWWA. The IWA/AWWA Water Audit Method provides water utilities with the necessary tools and guidance to manage their water supplies effectively, promoting efficient resource management practices. There is an array of tools and software that can be used to determine water loss in specific areas. For instance, the latest iteration of the AWWA Free Water Audit Software, version 6.0, released in 2020, is now accessible (American Water Works Association, n.d.). The software is widely recognised as the prevailing instrument within the industry for doing the yearly water audit and utilising the obtained findings to inform the development of a cost-efficient water loss management and revenue recovery programme. The initial release of this product occurred approximately 15 years ago, and it has since gained widespread adoption in North America and globally.

1.4.2 Pressure Management

Pressure management minimisation is an essential strategy for controlling water loss through leakages in WDNs by minimising waste and protecting the infrastructure. There is an associated correlation between water pressure and leakages in pipe volumes. Pressure management has, therefore, been used over time to control water leakages in many WDNs worldwide. Many models were identified for controlling pressure-related leakages in WDN. Pressure sensors effectively maintain the WDN's pressure with varying pressures by reducing or sustaining the water pressure in the valves. These sensors measure the pressure drop along the pipe for leakage control purposes. Pressure management can be implemented to suit the needs of the end users during low-demand periods for water conservation purposes (see Section 1.5.4 for further information).

1.4.3 Reliability Analysis

The reliability analysis (RA) of an WSS is an evaluation tool used to measure the ability of the system to deliver water to consumers consistently without interruption. The analysis targets the assessment of the performance and effectiveness of the WSS under different operating conditions. RA could be determined based on critical demand values (Surendran et al., 2005), cost optimisation, and water system component failure. Several steps are involved in keeping the RA of WSS effective using data gathered from the functioning of the components, such as the valve, pump, and pipes. For instance, data collection from pipeline evaluation is the most crucial element in assessing the dependability of water distribution systems (WDS). It can assess pipeline failure probabilities (Mazumder et al., 2019). The most significant and frequent type of burst failure has negative and adverse economic, environmental, and societal effects. The pipe's plastic collapse ultimate failure pressure, which shows the pipe's maximum load-bearing capability under internal pressure, is often used to express burst failure (Gao et al., 2019). According to Mazumder et al. (2021), machine learning techniques are more computationally efficient than the physics-based pipeline failure risk analysis method. Their work recommended that XGBoost has the optimal algorithm for predicting the failure of most pipelines. The placement of shut-off valves also significantly impacts system reliability since they immediately affect which system components may be isolated and subsequently define the available flow routes.

1.4.4 Pump Scheduling

Water utilities can significantly reduce operating expenses by scheduling pumps more efficiently and shifting energy use to low-tariff times. Several pump scheduling techniques exist to maximise water quality and energy costs simultaneously. Due to the complexity of water networks and the non-linear behaviour of water flow, many techniques have been used for the formulation of pump scheduling. However, some computational methods are more effective and faster in optimising energy costs and water quality than others. For instance, Abdallah and Kapelan (2019) developed an efficient computational method, iterative Extended Lexicographic Goal Programming, to optimise energy cost and improve the chlorine concentration in water with variable speed pumps (VSPs). To check the fulfilment of the non-convex requirements, determine the actual cost, and address a wide range of networks such as VSPs, non-linear programming has also computed a global numerical approach with numerous enhanced techniques (Bonvin et al., 2021). A recent model was formulated to cater to energy conservation constraints in a way that can be used with non-convex Mixed Integer Non-Linear Programming optimisation solvers (Cantu-Funes & Coelho, 2022). The simulation from the study showed less energy use during peak periods compared to previous works and has been tested and validated to become the new benchmark in the US pump scheduling field.

1.4.5 Optimal Design, Operational Planning, and Control

Efficient water management necessitates the consideration of optimal design, operational planning, and control in water control systems. The optimal design process entails the development of a water control system with predetermined criteria while minimising associated expenses. Operational planning entails formulating strategies and tactics to optimise a system's functioning, ensuring its efficiency. Optimal control refers to systematically monitoring and making necessary adjustments to the water system to ensure its proper functioning as intended.

According to projections, there is an anticipated 50% rise in energy consumption within the water sector by 2030 (Reis et al., 2023). The energy demand in a WSS is subject to various factors, including but not limited to topography, proximity to end users, water losses, equipment efficiency, and the extent of treatment (Reis et al., 2023). The challenges within this context encompass operational considerations, the rising costs of energy, and an absence of equilibrium among diverse stakeholders, such as water utilities, regulatory entities, and political establishments. There is currently a lack of research on applying optimisation techniques to complex water networks over extended periods. Various control strategies exist that can be employed within water control systems. One commonly employed strategy for intake pump stations is level-based control. Optimal control strategies may be employed to minimise expenses and enhance operational effectiveness. This study examines the involvement in demand response mechanisms, implementation of energy efficiency measures, and utilisation of renewable energy sources.

1.4.6 Water Quality Assessment

Water quality determination is critical for any drinking water to be accepted and fit for human and animal consumption. Unsafe drinking water causes waterborne diseases (cholera, dysentery, and diarrhoea) in society. Continuous and standard monitoring measures of water quality are essential components of a WSS to curb unhealthy contaminants, pollutants, and toxic substances from compromising human health. Water quality testing can aid in the identification of potential dangers and issues, such as contamination, corrosion, leakage, disinfection byproducts, and microbiological development. Water quality assessment can also inform drinking WSS-related decision-making and management, such as optimising treatment procedures, enhancing operational efficiency, adhering to legislation and standards, and communicating with stakeholders and consumers.

In WSSs, there have been many inventions for the accurate determination of the problems with water pollution and for creating effective measures involved in monitoring and assessing water quality. Physical, chemical, and biological characteristics are often used to measure water quality assessment. The Water Quality Index (WQI) is the most useful metric for measuring the quality of water sources for human use. WQI helps simplify the bulkiness of water quality parameters and information logically so that all stakeholders understand it (Tyagi et al., 2013).

Uddin et al. (2022) assessed coastal water quality at Cork Harbour, Ireland, using WQI models. The available water quality indicator monitoring data is converted into a single number to measure water quality for decision-making using mathematical methods. Both specialists and non-experts can easily understand the method's results and implement it in most cases because of its simplicity. According to Sivaranjani et al. (2015), the score is usually computed between 0 and 100 with five different water quality categories of excellent (90–100: excellent), good (80–90: healthy), medium (50–80: alarming) bad (25–50), and very bad (25–0: very severely affected).

To ensure that end consumers receive clean water, the WSS must routinely maintain water purification and sanitation requirements. Aside from computational methods, measuring heavy metal pollution by indices is one of the most crucial methods for determining the quality of drinking water, and heavy metal data can be used to determine the WQIs (Ghaderpoori et al., 2018). Continuous monitoring is necessary since there is always a chance that heavy metals and hazardous chemicals will contaminate water supplies. Heavy metals leaking through iron pipes in distribution systems and geological contamination of the water's source region are the leading causes of heavy metal contamination in drinking water.

1.4.7 Vulnerability Assessment

The term "vulnerability" refers to a property that makes a critical infrastructure susceptible to being destroyed or rendered unusable by a threat in its development, application, or management. In a WSS, vulnerability assessment is crucial for system design and operation. An ideal all-encompassing WSS comprises water sources, raw water transport pipelines, water treatment facilities, and WDNs. However, because most of them are accessible and diversified in terms of space, these components and subsystems offer the most potential for natural and human-related interactions (Karamouz et al., 2010). As a result of high pressure on the supply-demand of water to the end users, the WDN and infrastructure deteriorate and fail to perform over time. Several factors are responsible for the vulnerability of WDNs because of their dependence on other interconnecting systems and infrastructures. Numerous issues, including decreases in system efficiency and capacity brought on by pipe tuberculation, leaks, corrosion, and other issues, can impact the quality and amount of provided water. A WSS undergoes frequent pipe bursts and breaks, which are associated with degradation and ageing. The cost of resources and time to fix some of these breaks is enormous. Also, dangerous threats such as natural hazards and those induced by humans can impair the performance of a WSS (Karamouz et al., 2010).

Vulnerability assessment is essential for a better understanding of the WSS for enhanced decision-making support for planning and management, and various WSS vulnerability assessment techniques (Yazdani & Jeffrey, 2012). According to studies, qualitative, quantitative, semi-quantitative, deterministic, and conventional methods have been explored and applied over time to examine vulnerable assessments (Anandhi & Kannan, 2018). This new method can transform a theoretical idea (the vulnerability of water resources) into a practical framework, particularly in the context of altered precipitation and temperature to a lesser extent.

1.4.8 Rehabilitation

Rehabilitating a WSS brings it back to its original state or increases its functionality if it has deteriorated. Pipes, pumps, and valves, among other system components, may require maintenance or replacement. When a system is being rehabilitated, it is done to increase its dependability and efficiency with as few interruptions to service as possible. System rehabilitation aims to improve system reliability and efficiency to maximise service continuity while minimising disruption. According to the Water Environment Federation (WEF, 2017), the specific procedures involved in rehabilitating a WSS can vary depending on the system's condition and the project's goals.

The benefits of WSS rehabilitation include bettering the water quality for users, boosting the pipeline's hydraulic capacity, minimising leaks, lowering the frequency of repairs, lowering the risk of property damage, and enhancing reliability. Compared to open trench replacement, pipeline rehabilitation methods are usually less expensive and disruptive to the neighbourhood. Several factors must be considered while choosing a pipeline rehabilitation procedure to ensure the most favourable result. The selection of an appropriate technique for rehabilitation will depend on the specific issue(s) that require attention, such as leaks, structural deterioration, corrosion, or reduced flow capacity. Various types of challenges require different approaches or techniques that are more suitable. The critical examination of the costs associated with each rehabilitation strategy is crucial. In addition to the initial expenditure, this analysis also considers long-term variables, such as maintenance costs, operational expenditures, and the projected lifespan of the rehabilitated pipeline. Determining the most economically efficient option necessitates a comprehensive evaluation of the expenses associated with multiple alternatives.

It is imperative to consider the site's physical characteristics, including factors such as the accessibility of the pipeline, the composition of the soil, the presence of utilities or barriers, and any potential environmental concerns. Specific methodologies may be more suitable for densely populated metropolitan environments or regions characterised by fragile ecosystems.

The anticipated life-cycle performance of the pipeline necessitates a thorough evaluation of its durability, longevity, and overall operational effectiveness. The selected methodology should provide a reliable and long-lasting resolution that meets the requirements for the expected lifespan of the service.

Within the framework for the application of contemporary advances (which incorporates ICT [Information and Communication Technology]), the popularity of the term "smart" is growing. Speaking and writing regarding Smart City, Smart Homes, Smart Grid, and Smart Maintenance is increasing. Whenever we notice the phrase "smart grids", we refer to electrical grids. However, the word is increasingly used to describe larger industrialised systems, such as wastewater, water supply, and thermal systems. We can enhance the efficiency of the WDS to develop smart cities by utilising artificial intelligence (AI) methodologies (Rojek & Studzinski, 2019; Sánchez-Corcuera et al., 2019; Pietrucha-Urbanik et al., 2020). To mitigate the system's collapse threshold, it is imperative to regulate performance at the source and pumping locations, as well as to rejuvenate the entire system. The former process is computationally complex due to the necessity of employing sophisticated optimisation software, whereas the latter incurs significant expenses. Consequently, an alternative method involves the rapid identification and localisation of malfunctions, which, in addition to the specified hidden water intrusion, are often not immediately apparent as they rely significantly on seepage water infiltrating the Earth's surface through fissures in degraded pipelines (Rojek & Studzinski, 2019; Pietrucha-Urbanik et al., 2020).

Continual assessment of water quality is the most essential measure for assuring the supply of high-quality water that meets the regulatory standards of each country. These regulations typically stipulate that tracking is to be regularly conducted in areas representative of the condition of the water in the supply network, where devices could be installed to identify pollution occurrences. Nevertheless, the process for locating such sites is not indicated. Due to the high expenses and additional factors, devices must not be installed at each connection node. Multiple techniques have been devised throughout the globe to address the display position issue. However, no standard exists to pinpoint illustrative sensor positions before this point (Beata et al., 2020).

The WDS is the primary water source in urban areas in our increasingly developed world. An adequate water supply of both quantity and quality are essential. Traditional techniques for measuring aquatic variables are costly and cannot identify water functionality in actual time (Baanu & Babu, 2022). It is necessary to observe the WDS circumstances on-site. Telecommunications and information technology are crucial in continuously tracking and managing the WDS (Baanu & Babu, 2022). The smart water grid (SWG) is devised to surmount the shortcomings of conventional assessment and instantaneously track the WDS. The SWG is designed to monitor and regulate water usage, leakage, and modifications in water conditions within the WDS. Smart water level tracking is a practical and economically feasible approach to checking parameters about water quality (Baanu & Babu, 2022). Secure water distribution, treatment, administrative challenges, enjoyment, and public security depend on constant, immediate tracking for water resource management. An improved understanding of the elements influencing quality may be gained with less time, money, and manpower spent on traditional studies (Baanu & Babu, 2022).

A typical approach for pressure control in district-metered areas (DMAs) or zones is the installation of pressure-reducing units (Gao et al., 2022). Researchers have generally agreed upon four essential pressure management approaches using a pressure-reducing valve (PRV): constant outlet, period altered, flow modified, and critical pressure control (Gao et al., 2022). Numerous studies are devoted to deciding where pressure-reducing locations should be assigned, which PRVs should be used, and how pressure should be regulated after they are built. Appropriate pressure-lowering stations are often constructed by computerised modelling (Gao et al., 2022). Nevertheless, the approaches offered proved to be tried and true amid network uncertainty. The pressure-lowering stations constructed by the DMAs or areas might fail leading to unnecessary expenditures (Gao et al., 2022). As a result, it is preferable to construct interim pressure-reducing units for use in field tests to determine where and what method should be employed to build the ideal pressure-reducing unit (Gao et al., 2022). An additional makeshift pressure-reducing station must be built for every site practice, which wastes resources, results in water interruptions throughout the procedure, and may result in dissatisfied customers. Moreover, designing and constructing makeshift pressure reduction units for tests at numerous sites is time-consuming and costly. AI and SCADA (supervisory control and data acquisition) were widely used during the trial, giving rise to cognitive pressure-regulating technologies. It may collect accurate information on the network's operations without affecting customers' water use and provide pressure control. Most water utilities' day-to-day decisions will be made by computers shortly (Gao et al., 2022).

1.4.9 Water Security

According to the United Nations (UN), water security can be described as the ability of a population to ensure long-term access to enough water that meets acceptable quality standards. This is crucial for sustaining livelihoods, promoting human well-being, supporting socioeconomic development, protecting against water pollution and disasters, and preserving ecosystems. It is important to note that water security is contingent upon a peaceful and politically stable environment. It is imperative to secure water systems susceptible to harm, reduce the consequences of water-related disasters such as floods and droughts, ensure access to water functions and services, and manage water resources comprehensively and fairly. Pursuing universal water security by the SDGs has prompted a careful examination of various techniques (HLPW, 2018). However, achieving this goal necessitates the development of an integrated framework that can evaluate the worldwide feasibility of integrating green and grey infrastructures (Vörösmarty et al., 2021).

Countries that invest significantly in engineering to maintain water security while depleting natural capital will face substantial limitations in achieving effective green-grey water management. The circumstances are anticipated to be particularly constraining in the developing world, where economic, technical, and governance capabilities to address these challenges are still inadequate. Without effective stewardship, the costs associated with water resource infrastructure and services will likely increase significantly, hindering efforts towards universal and sustainable water security.

Vörösmarty et al. (2021) results demonstrate that policies that support blended green-grey approaches offer a pathway to future global water security but will require a strategic commitment to preserving natural capital. Without such stewardship, water resource infrastructure and services costs will likely rise substantially and frustrate universal and sustainable water security efforts.

1.5 WATER SUPPLY NETWORK MANAGEMENT APPLICATION DOMAIN

The domain of water supply network management application encompasses a wide range of topics related to the functioning and administration of water supply networks. The system covers many components, such as water quality monitoring (WQM), leak detection and monitoring, pressure control and monitoring, parameter estimation and monitoring (PEM), state estimate and monitoring, and demand prediction and monitoring. Nevertheless, integrating the cyber-physical system with WSS infrastructures is a technical paradigm shift that will improve the water system's operations (Adedeji & Hamam, 2020).

1.5.1 Leakage Detection and Monitoring

In our current age, the demand for usable water is perpetually increasing. Climate change has resulted in a breakdown in the water cycle, which has affected water availability (Adedeji et al., 2022). Consequently, the shortage of water is a significant global issue. As the worldwide water shortage crisis worsens, it has become crucial for the water sector to employ a technological transition system to effectively utilise water resources. Large-scale, complicated WSSs or WDNs are intended to provide safe water nationwide. The universal accessibility of clean, potable water adheres to the SDGs defined by the UN. Nevertheless, the realisation of this objective may be hampered by several obstacles to the effective functioning of urban WSSs. The regular disappearance of water in the system is a significant obstacle to the function of the WSS. Even though the volume of water depletion due to leakage pipelines differs from system to system, it occurs in every system.

This presents numerous additional difficulties for water utilities. More difficulties that could come up because of network expenses include the comparatively high expenses linked to emptying caused by leakage, a possible negative effect on water quality, a greater probability of water pollution, monetary consequences, and the higher price of upkeep for fixing faulty pipelines (Adedeji et al., 2022). When the WDS leaks, additional energy is required to maintain the intended quality of functioning (Adedeji et al., 2022). Tragically, the cost of power consumption associated with extra water extraction is a significant concern. Leakage is additionally spotted by identifying soil cavities generated by water leaking or by discovering areas in pipelines with differences in characteristics because of an increase in the dielectric qualities that result from nearby saturated soils. This approach is lengthy and costly. However, it can be applied to metals or rubber pipelines.

The free-swimming systems approach involves inserting tubes that include an integrated energy source, electronic elements, and apparatus (acoustic detector, accelerometer, magnetometer, GPS-synchronised sound sender, and thermometer reader) into the body of water. These compartments capture the pipe's internal surroundings and transmit its contents to a server. Examining recorded data authorises the recognition and localisation of leaking water oddities. This technique is well-suited for large-diameter pipelines (Mashhadi et al., 2021). Modern advances in the continuous tracking of water supply networks have created fresh possibilities to earn the development of data-based approaches to detecting and localising water leaks. Approaches centred on deep learning are frequently utilised to identify and pinpoint water leakages in WSSs. Artificially created Neural Networks (ANNs) can be used to locate and spot water leakage in WSSs. Employing a hydraulic simulation of the water supply network, data have been obtained for different situations and cases with varying quantities of water leakage. The technology accurately identified small-scale WSS breaches using the radial basis function neural network technique to detect leaks. The hydraulic application – Environmental Protection Agency Network Evaluation Tool (EPANET) – and water-quality modelling software developed by the U.S. Environmental Protection Agency (EPA) for simulating the movement and chemical quality of water within pressurised pipe networks.

The pressure and flow variations in the water network served as the ANN model's data source, while the leaks' severity and locations served as result variables. Research has demonstrated that this approach could identify the magnitude and location of a breach with 98% precision and utilised the application of ANN to locate abnormalities found in water distribution series information using a pattern-matching strategy. This methodology was based on research comparing novel events to existing profiles.

The quantities of water leaking vary according to the kind and severity of conduit abrasion, the frequency of the discharge, and the surrounding pressure at the affected region. In such circumstances, DMAs' distribution networks can assist with identifying leaks and tracking (Katarzyna et al., 2020) and leak suppression via pressure regulation. It necessitates developing classification programmes (classifiers) that enable the handling and evaluation of information about pressure sources in real time and via the Internet. Implementing classifiers could boost the precision of leakage detection (Katarzyna et al., 2020). The accuracy matrix data relating to point force and pipe movement can also be utilised in the WSS to identify leaks (Katarzyna et al., 2020). The application of the connection proportion matrix derived using the fault state and typical state proved successful in enhancing the occupancy rate of water systems (Katarzyna et al., 2020).

The inverse transient approach can also detect water leaks in WDSs by determining the location of leaks using observed transient data. A hydraulic model mimics the system's behaviour in this approach to reduce the variance between the simulated and real-world values. As a novel method, inverse transient analysis integrates the inverse technique with network transient analysis. The inverse approach for leakage identification was theoretically developed. However, it has problems, such as needing precise knowledge of friction factors in all pipes for leak locations. In addition, real-world networks seldom reach a steady state; therefore, steady-state analysis is unsuitable for constant surveillance.

The approach was shown to be effective for leak detection and network calibration by analysing brief fluctuations. This technique also enables the rapid localisation of unexpected breaks. The Levenberg-Marquardt approach for minimisation converges more quickly than the Newton-Raphson method. Other optimisation methods, such as genetic algorithms, are used for leakage detection and roughness calibration across broad search spaces.

1.5.2 Water Quality Monitoring

WQM involves regular assessment and analysis of various physical, chemical, and biological parameters in water bodies. The primary goal of WQM is to ensure the safety and health of aquatic ecosystems and human health when using water for drinking, recreational activities, or industrial purposes. This monitoring process provides essential data for decision-making, regulatory compliance, and identifying potential environmental issues. WQM involves the analysis of a wide range of parameters, including temperature, turbidity, colour, odour, total suspended solids, and other chemical parameters, including pH, dissolved oxygen, biochemical oxygen demand, chemical oxygen demand, nutrients (nitrogen and phosphorus), heavy metals, pesticides, and various organic and inorganic compounds. It also includes biological parameters, such as the presence of pathogens, indicator bacteria (e.g., *Escherichia coli*), and assessment of the aquatic ecosystem's biodiversity. WQM can occur at various locations, such as lakes, rivers, streams, ponds, and wastewater treatment plants. It can also ensure the quality of water extracted from wells used as drinking water supply and water quality in estuaries and nearshore zones (Zessner, 2021).

WQM can be performed through Grab Sampling, which involves collecting water samples at specific points and times for laboratory analysis. It can also be performed by using sensors and automated equipment to continuously measure parameters in real time and utilise satellite or aerial imagery to assess water quality in large water bodies. Environmental regulations and permits often mandate WQM. Regulatory authorities set standards and limits for specific parameters, and water utilities, industries, and other stakeholders must adhere to these guidelines. Monitoring water quality is crucial in conducting environmental impact assessments for various projects, such as industrial developments, mining activities, and construction projects.

Assessing potential impacts on water quality helps make informed decisions and implement mitigation measures (Kumar et al., 2022). It is also essential to ensure the safety of water supplies. Contaminated water can pose health risks, including waterborne diseases, which makes continuous monitoring crucial for public health protection. It can be integrated into early warning systems to promptly detect and respond to pollution incidents. Timely identification of pollution events can help prevent further contamination and mitigate impacts on ecosystems and human health. Water quality data collected over time can be analysed to identify long-term trends and changes in water quality. These trends can provide valuable insights into the effectiveness of pollution control measures and the overall health of water bodies (Kumar et al., 2022).

WQM is integral to environmental management and human health protection. By regularly assessing water quality parameters and responding to potential issues proactively, we can ensure the sustainable use and preservation of our valuable water resources for future generations. WQM generates vast amounts of data from multiple monitoring points and parameters (Kumar et al., 2022). Effective data management and analysis are essential for extracting valuable insights and making informed decisions. Advanced data analytics and visualisation tools help identify patterns, trends, and potential correlations between water quality parameters. It can exhibit seasonal variability influenced by rainfall, temperature, and human activities. Climate change can also impact water quality by altering precipitation patterns, leading to more intense storms and changes in water temperature. Continuous monitoring helps us understand these variations and their potential implications. International collaboration and data sharing between neighbouring countries are essential for effectively addressing transboundary water quality challenges. WQM is an integral part of Integrated Water Resources Management (IWRM), which takes a holistic approach to managing water resources sustainably. IWRM considers social, economic, and environmental factors to balance competing water demands while safeguarding water quality (Baanu & Babu, 2022).

WQM programmes can serve as educational tools to raise awareness among the public, policymakers, and industries about the importance of water conservation, pollution prevention, and sustainable water management. Non-point source pollution, caused by diffuse runoff from urban areas, agriculture, and construction sites, can significantly impact water quality. Monitoring and implementing the best management practices are essential to address this type of pollution. WQM is an ongoing and dynamic process that requires collaboration between various stakeholders, including government agencies, research institutions, water utilities, industries, and the public. By continually improving monitoring techniques, sharing knowledge, and implementing effective management strategies, we can safeguard our water resources for future generations and ensure a sustainable and healthy environment (Zessner, 2021).

Water quality samples must assess water's biological, chemical, physical, and microscopic properties. Large quantities of chemicals are utilised in everyday tasks, and the industry eventually finds its way into the water, making it extremely difficult to monitor water quality. In addition, there are currently few pristine water sources on the entire planet, and many types of water have been contaminated. Water-related contaminants profoundly impact human well-being and longevity. As globalisation increases, the whole globe is confronted with difficulties of water scarcity and pollution. In Palembang, South Sumatra, Indonesia, water storage facilities remain in utilisation, and a dearth of reservoir management and cleansing may cause the water in them to turn turbid. In the meantime, no mechanism is designed to avoid water discolouration or offer inhabitants a heads-up about water issues. The principles associated with Internet of Things (IoT) were utilised to serve occupants with immediate updates. We can determine its suitability for typical usage through an approach for measuring the turbidity, pH levels, temperature, conductance, and total dissolved solids (TDS) of water. TDS, turbidity, temperature, pH, and conductivity sensors are utilised. Every single sensor's knowledge is transmitted across the internet. Testing is conducted to obtain data, transmit digital information, track and document it, and analyse information about water health. The technology is now being used at the Engineering Building at the National Chin-Yi University of Technology in Taichung, Taiwan (Sung et al., 2021).

1.5.3 State Estimation and Monitoring

State estimation and monitoring are vital in numerous disciplines, such as engineering, management, and ecological science. The two concepts encompass estimating and forecasting a system's present or prospective condition using the data, findings, and computational models. State estimation, system recognition, or signal analysis is the inference or estimation of the internal condition factors of an unpredictable system by considering the system's data inputs and outputs. In simplified terms, it refers to the process that establishes a structure's present or preceding condition through measurements or observations. State estimation is utilised in various disciplines, including Management Devices, Tracking and Localisation, Weather Conditions Prediction, and Health Evaluation. State estimation is essential for feedback regulation for control technology (Khodarahmi & Maihami, 2023; Song & He, 2023).

By approximating the current condition of an evolving system, operators can arrive at sensible choices and modify the parameters to accomplish the targeted level of reliability and efficiency. State estimation is implemented in automation and independent systems for exact positioning and identification. A mobile automation system or vehicle can efficiently explore its surroundings by approximating its vantage point and perspective. State estimation is utilised in weather forecasting to predict adverse conditions using samples collected by different meteorological stations and orbiting satellites and computational weather programmes. In medical services, state estimation is often employed to track the well-being of individual patients by utilising sensor data from medical devices (Song & He, 2023).

State estimation and monitoring have become essential in the application area of water resources to administer water supply networks, effectively assuring water quality and tackling operational hurdles. The movement speeds and pressure levels at different points across the WDS are estimated using state estimation methods. State estimators can offer an in-depth understanding of the hydraulic function of a network by integrating flow meter, pressure sensor, and valve position sensor data. The data gathered aids in optimising water supply, detecting breaches, and guaranteeing consumers have appropriate pressure. Monitoring and state estimation are methods to identify and pinpoint WDN breaches. Defects indicative of potential breaches can be identified by analysing flow and pressure data (Vijayalakshmi et al., 2023).

It is very necessary to have an accurate forecast of water demand to maximise water transportation and storage. Water demand patterns may be estimated using state estimation techniques, including historical data, current weather situations, and other pertinent elements. This makes it easier for water utilities to properly manage their water supplies and prepare for future infrastructure demands that may arise. State estimation calibrates the hydraulic models that make up the WDS. Adjustments may be made to improve the framework's validity by juxtaposing the data seen with the predictions generated by the model. After calibrating the model, it is used for scenario analysis, emergency response planning, and system performance evaluation under various scenarios. Estimation and monitoring at the state level help locate potential sources of water waste, including but not limited to leaks, unauthorised usage, and erroneous meter readings. Utilities can adopt preventative actions to cut down on water losses, preserve resources, and increase the effectiveness of their networks by constantly tracking water flows and pressures (Gourisetti et al., 2023; Thebe, 2023).

Integrating state estimates and monitoring with modern data analytics enables the generation of real-time insights into operational processes. This enables operators to make educated choices about the operation of valves, the scheduling of pumps, and the control of pressure to maximise system performance and reduce energy usage. Identifying probable contamination occurrences or pollution in the water supplies is made possible via time and real-time monitoring of water quality and state estimates. For the sake of protecting public health and halting the spread of pollution, a prompt reaction to incidents of this kind is necessary. Techniques for estimating the state may also be used to evaluate the condition of vital components of the infrastructure used to provide water, such as pipes and pumps. Monitoring parameters like pressure changes and flow rates enables utilities to spot possible problems and plan maintenance, thereby reducing downtime for their systems and the number of interruptions they experience (Vijayalakshmi et al., 2023).

1.5.4 Pressure Management

Pressure management in water supply refers to controlling water pressure within the distribution network. It involves adjusting pressure levels at different points in the system to ensure water delivery at appropriate pressures while minimising water losses and energy consumption. Proper pressure management is essential for improving WSS' efficiency, reliability, and sustainability. Excessive pressure in the distribution network can lead to leakages and bursts, resulting in significant water losses. By managing pressure, utilities can minimise the occurrence of leaks and conserve water resources. High pressure demands more energy for pumping and distribution; pressure management helps and optimises pump operations, thereby reducing energy consumption and associated costs. Excessive pressure can also lead to increased stress on pipes, potentially shortening their lifespan; by maintaining optimal pressure levels, utilities can extend the life of the distribution infrastructure (Cen et al., 2023; Kraft, 2023; Lancioni et al., 2023).

A typical step that may be taken to accomplish pressure control for DMAs and zones involves the erection of pressure-reducing units. Constant location, time modification, circulation alteration, and essential pressure regulation are the four basic pressure management approaches that have become universally recognised by experts. Each of the techniques uses a PRV. Many research studies are devoted to establishing pressure-regulating techniques, deciding where to locate pressure-reducing areas, choosing PRVs, and generating location judgements to facilitate the construction of appropriate pressure-reducing stations (Bideris-Davos & Vovos, 2023).

PRVs are commonly used devices in pressure management. They are installed in the WDS to reduce the pressure from higher supply mains to lower-pressure zones or consumers. PRVs maintain consistent pressure within specific limits, preventing excessive pressure at the consumer end. Pressure-sustaining valves are used in areas with low demand or during low consumption periods. These valves maintain the minimum required pressure in the distribution network by releasing excess pressure and preventing negative pressure in the system. WDNs are often divided into pressure zones. A PRV controls each zone to maintain the desired pressure level. This zoning approach allows utilities to optimise pressure management based on demand patterns and network characteristics. Advancements in technology have enabled the implementation of smart pressure management systems. These systems use real-time data, including flow rates, demand patterns, and pressure measurements, to adjust pressure levels. Smart pressure management improves system efficiency and responsiveness. Pressure transients or surges can occur in WDNs due to sudden changes in flow or valve operations. Transient analysis helps understand and mitigate the impact of such events on the system to prevent damage and water quality issues (Gao et al., 2022; Bideris-Davos & Vovos, 2023).

Modelling in mathematics is a frequent method used in constructing appropriate pressure-reducing stations (Gao et al., 2022). On the other hand, the approaches offered did not undergo evaluation under conditions of uncertainty in the networks. Because of a dearth of certainty around DMAs or sections, the pressure-reducing stations they constructed could not adequately pressure management's actual demands, leading to a loss of capital (Gao et al., 2022). Because of this, constructing transitory pressure-reducing units to conduct field tests to determine where and what method should be employed to construct the most effective pressure-reducing unit is the more appropriate choice (Gao et al., 2022). Nonetheless, to conduct every area of the test, it is necessary to construct another transient pressure-reducing unit. This results in the loss of money and the interruption of the water service for the test, which may lead to discontented clients.

Furthermore, repeatedly planning and building transient pressure reduction units in various places to conduct tests proves extremely challenging. During the field study, a smart pressure-regulating innovation was developed as a direct consequence of the broad adoption of AI systems and SCADA (Gao et al., 2022). SCADA systems are often utilised to monitor and control WDNs. Integrating pressure management strategies with SCADA allows remote monitoring, data collection, and automated pressure adjustments based on system conditions. Public education about the importance of pressure management and its benefits is essential. Consumers should be aware of the acceptable pressure levels for their premises and encouraged to report any unusually high or low-pressure incidents. Pressure management strategies may vary during peak and off-peak hours. By adjusting pressure levels during periods of high demand, utilities can ensure adequate supply while optimising energy usage. Pressure management is a vital aspect of WSS operations. By implementing appropriate pressure management strategies, water utilities can reduce water losses, enhance energy efficiency, extend the life of distribution infrastructure, and deliver consistent and reliable water services to consumers.

This was accomplished as an outcome of the broad adoption of these two technologies. It can acquire information relating to the system's operation in real time and implement management of pressure without affecting the amount of water used by customers. In the foreseeable future, computers will make most day-to-day decisions for water corporations (Gao et al., 2022).

In recognition of the fact that there is a clear relationship connecting pressure with leakage rate, controlling the pressure in WDSs represents one of the most efficient measures that can be taken to limit the amount of water that is lost via leaks (García et al., 2019). Therefore, PRVs play a significant part in lowering the amount of lost water. Much research has been done on the most effective way to cut down on leaks, and one of those ways is figuring out where the optimum place is to put PRVs and how many of them there should be (García et al., 2019). Gupta used an optimisation technique to establish the ideal position of PRVs in a system. Consequently, there was a 21% decrease in the amount of leakage that occurred. Instantaneous control was used by García et al. (2019) to successfully regulate the pressure that was present in a PRV that was installed in a network intake. This made it possible to achieve ideal operating conditions by lowering the degree of pressure at the points and, as a result, minimising the leakages.

1.5.5 Parameter Estimation and Monitoring

PEM in the water supply application domain involve the estimation of key parameters related to WDSs and the continuous monitoring of these parameters to ensure an efficient and reliable supply of safe drinking water. Parameter estimation techniques estimate water flow rates at various points within the distribution network system. Accurate flow rate estimation helps optimise water distribution and identify potential leakages. Monitoring and parameter estimation are employed to detect and locate leaks in the water supply network (Hou et al., 2023; Moosavian et al., 2023).

By continuously monitoring flow rates, pressure levels, and other system parameters, anomalies indicative of potential leaks can be detected. Parameter estimation techniques, such as hydraulic modelling and anomaly detection algorithms, can further pinpoint the leak's location, facilitating prompt repairs. Parameter estimation techniques help estimate pressure levels at different points in the distribution network. Pressure estimation can be performed by analysing flow rates, pipe characteristics, and valve positions. Monitoring pressure continuously ensures that the pressure within the network is maintained within safe and optimal limits to avoid burst pipes and water losses (Moosavian et al., 2023).

Monitoring water quality parameters, such as pH, chlorine levels, turbidity, and microbial contaminants, is critical for ensuring a safe and potable water supply. Parameter estimation may predict water quality trends based on historical data and observations. Real-time monitoring triggers alarms for any sudden changes in water quality that may indicate contamination. Parameter estimation techniques help estimate water demand patterns based on historical data, population growth, weather conditions, and other factors. Utilities can optimise water distribution, storage, and supply strategies by accurately forecasting water demand. Parameter estimation can be used to estimate the water age within the distribution network, which is the time water travels from the source to the consumers. Monitoring water age is crucial to prevent water stagnation and ensure the supply of fresh and high-quality water to consumers (Hou et al., 2023).

PEM are applied to assess the condition of critical water supply infrastructure, such as pipelines and pumps. By monitoring parameters like pressure, flow rates, and vibrations, utilities can identify potential issues and schedule maintenance proactively, minimise disruptions, and extend asset life. PEM aid in identifying areas of water loss, such as leakages and unauthorised consumption. Utilities can implement proactive measures to reduce water losses and improve the system's efficiency by continuously monitoring water flows, pressures, and other parameters. Integrating parameter estimation and tracking data with advanced analytics and decision support systems provides real-time operational insights (Cemiloglu et al., 2023; Reis et al., 2023). This helps operators make informed decisions regarding pressure management, water distribution, and asset maintenance.

1.5.6 Pipe Health Monitoring

Pipe health monitoring in the water supply network management application domain involves continuous assessment and evaluation of the condition of pipes within the distribution system. The goal is to detect and identify potential issues, such as leaks, corrosion, and structural degradation, to proactively address them and ensure efficient and reliable water delivery to consumers. Pipe health monitoring systems use valves to detect leaks in the WDN (Rojek & Studzinski, 2019; Adedeji et al., 2022). Any abnormal flow rates or pressure fluctuations indicate potential leaks, and the system triggers alerts for prompt action. Monitoring pressure levels in pipes is essential to identify anomalies that may signify excessive pressure, which can lead to pipe failures or leaks. Additionally, low pressure can indicate leaks or issues with supply, helping operators prioritise maintenance activities (Adedeji et al., 2022).

Corrosion is common in water supply pipes, especially in older systems. Pipe health monitoring employs corrosion sensors to detect the extent of pipe deterioration and corrosion rates, thereby allowing utilities to plan maintenance and replacement. Monitoring water quality within pipes is vital to ensure the water remains safe and potable during distribution. Sensors can detect changes in water quality, including chemical and microbial contaminants, which may result from pipe degradation or contamination events. Monitoring the flow rates within pipes helps understand water demand patterns and identify deviations that may indicate pipe blockages or flow restrictions. Advancements in technology have led to the development of smart pipe solutions equipped with sensors and communication capabilities. These pipes can monitor their condition, such as pressure, temperature, and structural integrity, providing real-time data for analysis (Damaševičius et al., 2023; Joseph et al., 2023).

Robotic technologies are increasingly used to inspect pipes from within, enabling the non-destructive assessment of pipe conditions. These robots can travel through pipelines, collect data, and identify potential issues without costly excavation. Pipe health monitoring data is integrated with asset management systems, allowing utilities to track the condition of their pipe infrastructure over time. This data-driven approach helps utilities schedule maintenance, repair, and replacement activities based on the condition and priority of pipes. By analysing historical monitoring data, advanced analytics can predict the future condition of pipes. Predictive models can estimate the remaining useful life of pipes and help prioritise replacements and maintenance (Damaševičius et al., 2023; Quamar et al., 2023).

Pipe health monitoring systems often incorporate data visualisation and decision support tools to provide water utility operators and engineers with actionable insights. These tools aid in understanding the state of the distribution system and assist in making informed decisions about maintenance and infrastructure investments. Pipe health monitoring is an essential aspect of water supply network management. By continuously monitoring the condition of pipes and utilising advanced technologies and data analytics, utilities can optimise maintenance efforts, reduce water losses, ensure water quality, and enhance the overall resilience and efficiency of the WDS (Joseph et al., 2023; Quamar et al., 2023).

1.5.7 Pump Energy Consumption Monitoring

Pump energy consumption monitoring in the water supply network management application domain involves continuous measurement and analysis of the energy usage of pumps within the distribution system. Pumps are critical components of water supply networks as they move water from the source to consumers. Monitoring pump energy consumption is essential for optimising energy efficiency, reducing operational costs, and ensuring a sustainable water supply. Advancements in pump technology have led to the development of smart pumps equipped with energy-monitoring capabilities. These pumps come with built-in energy meters and communication systems that provide real-time data on energy consumption. In cases where pumps are not equipped with built-in energy monitoring, external energy meters and sensors can be installed to measure pump energy consumption. These meters can be connected to a centralised monitoring system for data collection and analysis (Oh et al., 2023; Perelman & Fishbain, 2022).

Pump energy consumption data is logged and transmitted to a central monitoring system through telemetry or SCADA systems. This allows for remote monitoring and data analysis in real time. By analysing pump energy consumption data, water utilities can evaluate the energy efficiency of their pump operations. This analysis helps identify inefficiencies and areas for improvement. Load profile analysis involves studying the energy consumption pattern of pumps over time. Utilities can identify peak energy demand periods and adjust pump scheduling to reduce energy consumption during high-demand hours. Monitoring pump energy consumption helps assess pump performance. Deviations in energy usage from expected levels may indicate pump inefficiencies or mechanical issues that require attention. Comparing pump energy consumption with benchmark data and industry standards allows utilities to gauge their performance relative to similar systems. Benchmarking helps identify areas where energy conservation can be achieved (Zhou et al., 2023).

Pump energy consumption monitoring enables utilities to develop energy optimisation strategies. These strategies may include variable speed drives, pump scheduling, and efficiency improvements to reduce energy use. For large water utilities, pump energy consumption monitoring can aid in allocating energy costs accurately among different service areas or consumers. Implementing real-time alarms and alerts based on pump energy consumption thresholds helps quickly identify abnormal energy usage or potential equipment failures. Integrating pump energy consumption data with demand forecasting models helps utilities predict energy requirements and plan pump operations accordingly. Pump energy consumption monitoring supports the implementation of energy efficiency initiatives and the adherence to environmental regulations and sustainability goals. By continually monitoring pump energy consumption and implementing energy-saving strategies, water utilities can reduce operational costs, conserve energy resources, and enhance the overall sustainability of WSSs. Additionally, optimising pump energy consumption contributes to reduced greenhouse gas emissions and supports efforts towards a more environmentally friendly water supply network (Perelman & Fishbain, 2022; Oh et al., 2023; Zhou et al., 2023).

1.6 THE NEED FOR INDUSTRY 4.0 IN THE WATER SUPPLY NETWORK MANAGEMENT

Implementing Industry 4.0 in water supply network management offers significant advantages and addresses critical challenges water utilities face. Industry 4.0 integrates digital technologies and smart solutions to create a more interconnected, automated, data-driven environment. Real-time monitoring systems are gaining popularity in water supply network management. These systems continuously collect data from various sensors and in real time to detect leaks promptly. When a leak is detected, the system can generate alerts or notifications, allowing utilities to take immediate action. DMAs are the water supply network subdivisions that enable better water flow control and monitoring. By dividing the network into smaller zones, utilities can more accurately identify leaks within specific areas (Javaid et al., 2022; Lowe et al., 2022; Unhelkar et al., 2022).

Industry 4.0 technologies, such as IoT sensors and smart meters, enable real-time monitoring of WDNs. This data-driven approach helps optimise water flow, pressure, and demand patterns, leading to efficient water distribution and reduced water losses. IoT sensors and advanced analytics facilitate early leak detection, helping utilities promptly identify and repair leaks in the distribution system. This proactive approach reduces water losses and ensures the sustainable use of water resources. Water utilities can implement predictive maintenance strategies by leveraging data from sensors and IoT devices. Predictive analytics help identify potential issues with pumps, valves, and other infrastructure components, allowing utilities to schedule maintenance before significant failures occur (Yao et al., 2022).

Industry 4.0 technologies enable better energy management through real-time monitoring and control of pumps and other equipment. Water utilities can save substantially by optimising pump operations, and reducing energy wastage. IoT-based WQM systems provide continuous data on key parameters. This helps utilities detect changes in water quality and respond to potential contamination events more quickly, safeguarding public health. Industry 4.0 enables remote monitoring and control of WSSs, reducing the need for physical inspections and site visits. This remote capability improves operational efficiency and reduces labour costs. Industry 4.0 empowers water utilities with vast amounts of data from various sources. Advanced data analytics and AI tools enable utilities to make informed decisions, optimise operations, and plan based on real-time insights. Smart metering and remote monitoring enable water utilities to provide more accurate billing information and respond faster to customer inquiries. Enhanced customer service increases customer satisfaction and trust (Mourtzis et al., 2022).

Industry 4.0 technologies enhance the resilience of WSSs during natural disasters or emergencies. Real-time data on system conditions prioritise restoration efforts. By optimising water usage, reducing losses, and adopting energy-efficient practices, Industry 4.0 contributes to the sustainable management of water resources, supporting environmental conservation efforts. In conclusion, Industry 4.0 is pivotal in transforming water supply network management by leveraging digital technologies and data-driven approaches. Implementing smart solutions enhances efficiency, reduces costs, improves water quality, and ensures the reliable and sustainable delivery of safe drinking water to communities. Embracing Industry 4.0 in water supply network management is essential for meeting the challenges of the future and achieving a more resilient, efficient, and sustainable water supply infrastructure (Lowe et al., 2022; Mourtzis et al., 2022; Yao et al., 2022).

1.7 CHAPTER SUMMARY AND CONCLUSION

The WSS is a critical infrastructure that encompasses collecting, transmitting, treating, storing, and distributing water to various sectors, including residential, commercial, industrial, and agricultural, and fulfilling public requirements, such as firefighting and street cleaning. The provision of this service is of utmost importance within the realm of municipal services.

Enhancing water systems with AI tools can yield significant advantages in environmental conservation, public health promotion, and economic prosperity. The benefits of a well-managed WSS encompass enhancing the quality of water resources and preserving aquatic ecosystems and drinking water resources. It is a medium of alleviating drought impacts, averting waterborne infectious illnesses in the community, and mitigating the hazards associated with unintentional submersion during leisure activities. An improved and well-maintained WSS can enhance the economic development of nations by making a significant contribution towards eliminating poverty. Integrated approaches and measures should be engaged in minimising water use in agriculture, manufacturing, and power generation aimed at lesser wastage. Mitigating chemical waste contamination and natural pollution, such as acidification, is paramount to attaining secure water for society.

REFERENCES

Abdallah, M., & Kapelan, Z. (2019). Fast pump scheduling method for optimum energy cost and water quality in water distribution networks with fixed and variable speed pumps. *Journal of Water Resources Planning and Management*, *145*(12), 04019055.

Abebe, T. A., & Tucho, G. T. (2021). The impact of access to water supply and sanitation on the prevalence of active trachoma in Ethiopia: A systematic review and meta-analysis. *PLoS Neglected Tropical Diseases*, *15*(9), e0009644. https://doi.org/10.1371/journal.pntd.0009644

Adedeji, K. B., & Hamam, Y. (2020). Cyber-physical systems for water supply network management: Basics, challenges, and roadmap. *Sustainability*, *12*(22), 9555.

Adedeji, K. B., Ponnle, A. A., Abu-Mahfouz, A. M., & Kurien, A. M. (2022). Towards digitalization of water supply systems for sustainable smart city development—Water 4.0. *Applied Sciences*, *12*(18), 9174.

American Water Works Association (n.d.). https://www.awwa.org/Resources

Anandhi, A., & Kannan, N. (2018). Vulnerability assessment of water resources – Translating a theoretical concept to an operational framework using systems thinking approach in a changing climate: Case study in Ogallala Aquifer. *Journal of Hydrology*, *557*, 460–474.

Baanu, B. B., & Babu, K. J. (2022). Smart water grid: A review and a suggestion for water quality monitoring. *Water Supply*, *22*, 1434–1444.

Beata, K., Dariusz, K., & Ewa, H. (2020). Fractal-heuristic method of water quality sensor locations in the water supply network. *Water*, *12*(3), 832.

Bello, O., Abu-Mahfouz, A. M., Hamam, Y., Page, P. R., Adedeji, K. B., & Piller, O. (2019). Solving management problems in water distribution networks: A survey of approaches and mathematical models. *Water*, *11*(3), 562.

Bideris-Davos, A. A., & Vovos, P. N. (2023). Algorithm for appropriate design of hydroelectric turbines as replacements for pressure reduction valves in water distribution systems. *Water*, *15*(3), 554.

Bonvin, G., Demassey, S., & Lodi, A. (2021). Pump scheduling in drinking water distribution networks with an LP/NLP-based branch and bound. *Optimization and Engineering*, *22*, 1275–1313.

Cantu-Funes, R., & Coelho, L. C. (2022). Simulation-based optimization of pump scheduling for drinking water distribution systems. *Engineering Optimization*. https://doi.org/10.1080/0305215X.2022.2044030

Cemiloglu, A., Licai, Z., Ugurenver, A., & Nanehkaran, Y. A. (2023). Optimal exploitation of urban water supply networks based on pressure management with the nondominated sorting differential evolution (NSDE) algorithm. *Water*, *15*(14), 2583.

Cen, H., Huang, D., Liu, Q., Zong, Z., & Tang, A. (2023). Application research on risk assessment of municipal pipeline network based on random forest machine learning algorithm. *Water*, *15*(10), 1964.

Damaševičius, R., Bacanin, N., & Misra, S. (2023). From sensors to safety: Internet of Emergency Services (IoES) for emergency response and disaster management. *Journal of Sensor and Actuator Networks*, *12*(3), 41.

Gao, J., Li, K., Wu, W., Chen, J., Zhang, T., Deng, L., & Xin, P (2022). Innovative water supply network pressure management method—The establishment and application of the intelligent pressure-regulating vehicle. *Energies*, 15, 1870.

Gao, J., Yang, P., Li, X., Zhou, J., & Liu, J. (2019). Analytical prediction of failure pressure for pipeline with 534 long corrosion defect. *Ocean Engineering*, *191*, 106497.

García, I. F. Novara, D., & McNabola, A. (2019). A model for selecting the most cost-effective pressure control device for more sustainable water supply networks. *Water*, *11*(6), 1297–1297. https://doi.org/10.3390/w11061297

Ghaderpoori, M., Jafari, A., Ghaderpoury, A., & Karami, M. (2018). Heavy metals analysis and quality assessment in drinking water–Khorramabad city, Iran. *Data in Brief*, *16*, 685–692.

Gourisetti, S. N. G., Bhadra, S., Sebastian-Cardenas, D. J., Touhiduzzaman, M., & Ahmed, O. (2023). A theoretical open architecture framework and technology stack for digital twins in energy sector applications. *Energies*, *16*(13), 4853.

HLPW, U. (2018). *Making Every Drop Count: An Agenda for Water Action*. JLPW Outcome Report.

Hou, S., Tao, T., Wang, C., & Zong, H. (2023). Risk assessment and leakage prediction system of the water distribution system of Changzhou, China. *Water International*, *48*(3), 344–357.

Javaid, M., Haleem, A., Singh, R. P., Suman, R., & Gonzalez, E. S. (2022). Understanding the adoption of Industry 4.0 technologies in improving environmental sustainability. *Sustainable Operations and Computers*, *3*, 203–217.

Joseph, K., Sharma, A. K., van Staden, R., Wasantha, P. L. P., Cotton, J., & Small, S. (2023). Application of software and hardware-based technologies in leaks and burst detection in water pipe networks: A literature review. *Water*, *15*(11), 2046.

Karamouz, M., Saadati, S., & Ahmadi, A. (2010). Vulnerability assessment and risk reduction of water supply systems. In R. N. Palmer (Ed.), *World Environmental and Water Resources Congress 2010: Challenges of Change* (pp. 4414–4426). ASCE (American Society of Civil Engineers).

Khodarahmi, M., & Maihami, V. (2023). A review of Kalman filter models. *Archives of Computational Methods in Engineering*, *30*(1), 727–774.

Kraft IV, J. J. (2023). Alternative approach for assessment of hydraulic design basis for pressure pipe liners. (Doctoral dissertation, Louisiana Tech University). Louisiana Tech Digital Commons. https://digitalcommons.latech.edu/dissertations/989/

Kumar, A., Palmate, S. S., & Shukla, R. (2022). Water quality modelling, monitoring, and mitigation. *Applied Sciences*, *12*, 11403.

Lancioni, N., Parlapiano, M., Sgroi, M., Giorgi, L., Fusi, V., Darvini, G., … & Fatone, F. (2023). Polyethylene pipes exposed to chlorine dioxide in drinking water supply system: A critical review of degradation mechanisms and accelerated aging methods. *Water Research*, *238*, 120030.

Lowe, M., Qin, R., & Mao, X. (2022). A review on machine learning, artificial intelligence, and smart technology in water treatment and monitoring. *Water*, *14*(9), 1384.

Mashhadi, N., Shahrour, I., Attoue, N., El Khattabi, J., & Aljer, A. (2021). Use of machine learning for leak detection and localization in water distribution systems. *Smart Cities*, *4*, 1293–1315.

Mazumder, R. K., Salman, A. M., Li, Y., & Yu, X. (2019). Reliability analysis of water distribution systems using physical probabilistic pipe failure method. *Journal of Water Resources Planning and Management*, *145*(2), 04018097.

Moosavian, N., Kasaei, M., & Roodsari, B. K. (2023). Using adaptive neuro-fuzzy inference system and imperialist competitive algorithm for leak detection in pipe networks. *Measurement*, *220*, 113336.

Mourtzis, D., Angelopoulos, J., & Panopoulos, N. (2022). A literature review of the challenges and opportunities of the transition from Industry 4.0 to Society 5.0. *Energies*, *15*(17), 6276.

Oh, H., Guk, I., Chung, S., & Lee, Y. (2023). Energy saving through modifications of the parallel pump schedule at a pumping station: A case study. *Journal of Water Process Engineering*, *54*, 104035.

Perelman, G., & Fishbain, B. (2022). Critical elements analysis of water supply systems to improve energy efficiency in failure scenarios. *Water Resources Management*, *36*(10), 3797–3811.

Pietrucha-Urbanik, K., Tchórzewska-Cieślak, B., & Eid, M. (2020). Water network-failure data assessment. *Energies*, *13*(11), 2990.

Quamar, M. M., Khan, K. A., & Khalid, M. (2023, May). Narrowband-IoT based integrated framework for monitoring pipeline condition in oil and gas industry. In *2023 International Conference on Control, Automation and Diagnosis (ICCAD)* (pp. 1–6). IEEE.

Reis, A. L., Lopes, M. A., Andrade-Campos, A., & Antunes, C. H. (2023). A review of operational control strategies in water supply systems for energy and cost efficiency. *Renewable and Sustainable Energy Reviews*, *175*, 113140.

Rojek, I., & Studzinski, J. (2019). Detection and localization of water leaks in water nets supported by an ICT system with artificial intelligence methods as a way forward for smart cities. *Sustainability*, *11*(2), 518.

Sánchez-Corcuera, R., Nuñez-Marcos, A., Sesma-Solance, J., Bilbao-Jayo, A., Mulero, R., Zulaika, U., Azkune, G., & Almeida, A. (2019). Smart cities survey: Technologies, application domains and challenges for the cities of the future. *International Journal of Distributed Sensor Networks*, *15*(6), 1550147719853984.

Sivaranjani, S., Rakshit, A., & Singh, S. (2015). Water quality assessment with water quality indices. *International Journal of Bioresource Science*, *2*(2), 85–94.

Song, J., & He, X. (2023). Robust state estimation and fault detection for Autonomous Underwater Vehicles considering hydrodynamic effects. *Control Engineering Practice*, *135*, 105497.

Stringer, L. C., Mirzabaev, A., & Benjaminsen, T. A., et al. (2021). Climate change impacts on water security in global drylands. *One Earth*, *4*, 851–864. https://doi.org/10.1016/j.oneear.2021.05.010

Sung, W.-T., Fadillah, F. N., & Hsiao, S.-J. (2021). IoT-based water quality monitoring. *Sensors and Materials*, *33*(8), 2971–2971.

Surendran, S., Tanyimboh, T. T., & Tabesh, M. (2005). Peaking demand factor-based reliability analysis of water distribution systems. *Advances in Engineering Software*, *36*(11–12), 789–796.

Thebe, T. (2023). SWAT runoff modeling and salinity estimation in the Odra River Catchment. (Master's thesis, Stockholm University, Department of Physical Geography). DiVA Portal. https://urn.kb.se/resolve?urn=urn:nbn:se:su:diva-219145

Thornton, J., Sturm, R., & Kunkel, G. (2008). *Water Loss Control*. McGraw-Hill Education.

Tyagi, S., Sharma, B., Singh, P., & Dobhal, R. (2013). Water quality assessment in terms of water quality index. *American Journal of Water Resources*, *1*(3), 34–38.

Tzanakakis, V. A., Paranychianakis, N. V., & Angelakis, A. N. (2020). Water supply and water scarcity. *Water*, *12*(9), 2347.

Uddin, M. G., Nash, S., Rahman, A., & Olbert, A. I. (2022). A comprehensive method for improvement of water quality index (WQI) models for coastal water quality assessment. *Water Research*, *219*, 118532.

UN World Water Development Report. (2020). Water and Climate Change. https://www.unwater.org/publications/un-world-water-development-report-2020

Unhelkar, B., Joshi, S., Sharma, M., Prakash, S., Mani, A. K., & Prasad, M. (2022). Enhancing supply chain performance using RFID technology and decision support systems in the industry 4.0–A systematic literature review. *International Journal of Information Management Data Insights*, *2*(2), 100084.

Van Duuren, F. A., Koschade, W., & Chapman, H. C. (1980). *A Guide to the Control of Water Losses in Pipe Networks*. Water Research Commission.

Vijayalakshmi, S., Savita, & Durgadevi, P. (2023). AI and IoT in improving resilience of smart energy infrastructure. In B. Balusamy & R. K. Dhanaraj (Eds.), *AI-Powered IoT in the Energy Industry: Digital Technology and Sustainable Energy Systems* (pp. 189–213). Cham: Springer International Publishing.

Vörösmarty, C. J., Stewart-Koster, B., Green, P. A., Boone, E. L., Flörke, M., Fischer, G., Wiberg, D. A., Bunn, S. E., Bhaduri, A., McIntyre, P. B., Sadoff, C., Liu, H., & Stifel, D. (2021). A green-gray path to global water security and sustainable infrastructure. *Global Environmental Change*, *70*, 102344.

Water Environment Federation (WEF). (2017). Sanitary Sewer Rehabilitation Fact Sheet (WSEC-2017-FS-009). Collection Systems Committee. Alexandria, VA: Water Environment Federation.

Yao, Y., Wang, C., Na, J., Hossain, M. S. A., Yan, X., Zhang, H., … & Li, J. (2022). Macroscopic MOF architectures: Effective strategies for practical application in water treatment. *Small*, *18*(8), 2104387.

Yazdani, A., & Jeffrey, P. (2012). Water distribution system vulnerability analysis using weighted and directed network models. *Water Resources Research*, *48*(6), W06517.

Zessner, M. (2021). Monitoring, modeling and management of water quality. *Water*, *13*, 1523. https://doi.org/10.3390/w13111523

Zhou, S. W., Guo, S. S., Xu, W. X., Du, B. G., Liang, J. Y., Wang, L., & Li, Y. B. (2023). Digital twin-based pump station dynamic scheduling for energy-saving optimization in water supply system. *Water Resources Management*, *38*(8), 2773–2789.

2 Introduction to Industry 4.0

2.1 INTRODUCTION

The idea of Industry 4.0, born out of a need to improve on the previous revolution, was conceived in the 1990s. It was birthed eventually in 2011 during the Hanover exhibition in Germany (Bigliardi et al., 2023). Since then, the revolution has become evident, and many have adopted it. Industry 4.0 has been described using renditions like Industry 14.0, the Industrial Revolution, and the Smart Revolution. Industry 4.0 has significantly improved in technological fields, and scholars have used it in countries like the United States, Italy, and the United Kingdom. The scope of its adoption was packaged in the national plan of some nations and regarded as "Industrie du Futur" and the "National Plan Industry 4.0", respectively. To effectively articulate this development, the United Kingdom explained it as "high-value manufacturing", made possible using accrediting technologies (Kagermann et al., 2013). One of the fascinating implications of Industry 4.0 is that it has sufficient space to measure, analyse, and retain data operatively and productively (Pandey et al., 2023; Psarommatis et al., 2023).

Industry 4.0 explains how digital technologies are changing the manufacturing and industrial sectors. It is a change from traditional, manual ways of making things to a highly linked and automatic system. Industry 4.0 includes several new tools that are changing the way businesses work. The Internet of Things (IoT), artificial intelligence (AI), robots, big data analytics, cloud computing, digital manufacturing (3D printing), and cyber-physical systems (CPS) are all examples of these technologies. These improvements make it possible to build "smart factories" where tools, gadgets, and systems can relate to each other and work together on their own. These "smart factories" are more efficient, flexible, and environmentally friendly. It aims to simplify production processes, make the best use of resources, improve product quality, cut costs, and make businesses more competitive. Industry 4.0 could also change how supply lines, transportation, and business models work, with significant social and economic effects. Monitoring the distribution of goods, data distribution, and enhanced transportation handling can be encouraged by practising Industry 4.0 (Javaid et al., 2022; Rahman et al., 2022). According to Bigliardi et al. (2023), two main classes of Industry 4.0 are recognised, which are information technology and advanced human-machine interface.

Technology is essential in establishing Industry 4.0 within producing enterprises (Kagermann et al., 2013). Industry 4.0 amends and improves new technologies and policies. "Industry 4.0" refers to bringing together new technological advancements. It can also bring solutions with the use of a machine's connection. The technologies of Industry 4.0 not only allow businesses to link but also the stakeholders, turning

DOI: 10.1201/9781003633839-2

them into great fortune. By gaining information between products and users, Industry 4.0 has the prospect of creating better goods for society. Industry 4.0 solves restrictions by integrating and uniforming power sources into a centralised energy source (Pandey et al., 2023).

Another advantage of Industry 4.0 is that it encourages interoperability between different parts and systems used in production. Through this, machines, systems, and people may effectively communicate and collaborate throughout the value chain. Predictive maintenance is made possible by the IoT and data analytics in Industry 4.0. Continuous monitoring of machines and equipment allows for the early detection of impending faults or maintenance requirements, enabling prompt servicing or replacement (Javaid et al., 2022; Pandey et al., 2023). This method is helpful for cutting down on maintenance expenses and maximising time. Better management of resources and increased energy efficiency are two outcomes of Industry 4.0 technology. Manufacturers may save money and help the environment by collecting and analysing data to find waste, optimise operations, and cut energy use. Industry 4.0 allows for more adaptable and quick production methods, known as "agile manufacturing". Manufacturers can swiftly launch new goods, customise orders, and effectively manage small batch sizes if they can quickly change production lines, reconfigure machinery, and react to changing market needs. Changes in the workforce are inevitable because of Industry 4.0. The emphasis on human labour will shift from regular jobs that can be automated to those that need higher levels of thought, such as creative problem-solving and critical analysis. To adapt to the new requirements of the digital era, workers must acquire new skills and update existing ones (Pandey et al., 2023; Psarommatis et al., 2023).

Productivity and quality have increased because of Industry 4.0's emphasis on collecting and analysing data in real time and optimising production processes. Intelligent systems' early detection of flaws or variances helps reduce waste and maintain high-quality standards throughout manufacturing. Supply chain visibility and coordination are optimised with the help of Industry 4.0 technologies. The procurement of raw materials, production, transportation, distribution, just-in-time manufacture, and delivery are made possible through real-time data exchange and analysis. Industry 4.0 paves the way for novel business models and sources of income. Manufacturers may use connection and data to deliver value-added services and preserve long-term client relationships, for instance, by shifting from selling items to providing product-as-a-service or subscription-based models. It is worth noting that the rate of adoption of Industry 4.0 may vary from sector to sector and from company to company. This revolutionary change in manufacturing is predicated on the fundamental concepts of connection, digitisation, and automation (Javaid et al., 2022; Psarommatis et al., 2023).

Industry 4.0 has been an essential tool used globally in various sectors. Its applications vary significantly from manufacturing to logistics, supply chain management, energy utilities, agriculture, healthcare, transportation and automotive, aerospace and defence, e-commerce, financial services, construction, etc. One of its uses is in the agricultural food industry. This industry strives with various new and original innovations and uses them in production and many areas. Many companies have adopted Industry 4.0, which has incurred huge profits, generating much suspicion among the industries (Javaid et al., 2022; Pandey et al., 2023). The result in favourable new power production could be traced to Industry 4.0 because Industry 4.0 technologies can increase the system that distributes electricity in various ways; because of the improvement in technology, the trade of reviving energy sources has advanced in recent years.

Despite the immense usefulness and applications of Industry 4.0, it is facing some challenges in reinforcing and training personnel in digital technology (Birkel et al., 2019), the need for modern management skills, and the hardness in revamping organisational culture (Ferreira et al., 2023). The scientific committee recently had significant zeal in questioning the usefulness of Industry 4.0 to sustain the environment, economy, and society (Oztemel & Gursev, 2020). The technological complexity of Industry 4.0 stems from the need to incorporate and use a wide range of cutting-edge tools and methods. Compatibility, interoperability difficulties, and the complexity of managing various technologies may be difficult to solve. Data security and confidentiality are paramount in this age of ubiquitous digital connection and constant information flow. Data security, data integrity, and compliance with privacy legislation are all continuing issues in the context of Industry 4.0. A competent and flexible workforce is essential for the changes brought forth by Industry 4.0. Organisations face formidable obstacles in the form of upskilling and reskilling personnel to work with cutting-edge technology, managing the transfer of workers whose activities may be automated, and addressing the possible effect on employment (Javaid et al., 2022; Pandey et al., 2023).

Industry 4.0 technologies cannot be implemented without a solid foundation of smooth data sharing and instantaneous decision-making. Sufficient connections, high-speed networks, and dependable communication technologies are essential. However, the broad adoption of Industry 4.0 may be hampered by infrastructural shortages and low connections in certain areas. Implementing Industry 4.0 technologies may need substantial expenditures in infrastructure, equipment, and qualified staff, which may add up to a hefty price tag. There may be a digital gap between big and small businesses because of the expense of adoption and the possible lack of investment for smaller organisations or those with fewer resources. Managing organisational culture change is essential for many businesses adopting Industry 4.0. Resistance to change must be overcome, organisational silos must be broken down, and a culture of creativity, cooperation, and lifelong learning must be fostered, all of which make change in management vital. In some instances, the creation of legal frameworks lags the rapidity with which Industry 4.0 technologies evolve. Data protection, liability, intellectual property rights, and ethical concerns are only some challenges that need to be addressed with well-defined legislation and standards in developing technology (Javaid et al., 2022).

Interoperability and scalability issues might arise when taking an Industry 4.0 project from the prototype stage to total production. Gaining the full advantages of Industry 4.0 requires ensuring interoperability and scalability across systems, equipment, and processes, particularly in intricate supply chains. While Industry 4.0 can potentially improve productivity and resource allocation, it raises questions about its long-term sustainability and potential adverse environmental effects (Javaid et al., 2022; Psarommatis et al., 2023). In the context of Industry 4.0, it might be challenging to balance economic development and sustainability concerns, such as energy consumption, e-waste management, and ecological footprints. Industry players, governments, academic institutions, and regulatory agencies must all work together to find solutions to these problems. Fostering an environment that encourages the creation and spread of Industry 4.0 technologies requires the implementation of regulations and funding for R&D, education, and the promotion of digital literacy. Today's industrial businesses, with growing management systems for better control of their invention, process, and supply, are part of the Industry 4.0 invention. The amounts of data that are generated using Industry 4.0 technologies are challenging for businesses to manage and analyse, and the usage of Industry 4.0 might be difficult and problematic for small-scale businesses due to a lack of resources and experience (Pandey et al., 2023; Psarommatis et al., 2023).

A brief journey into the focus of the evolutionary trend of industrial development revealed that the very First Industrial Revolution (Industry 1.0) incorporated steam- and water-powered mechanical production systems. The Second Industrial Revolution (Industry 2.0) was named after electrical energy-based mass production. The implementation of microelectronic technology and automation in manufacturing surfaced during the Third Industrial Revolution (Industry 3.0).

Industry 4.0 is backed up by nine technological pillars: big data, the cloud, Industrial Internet, simulation, vertical/horizontal integration, additive manufacturing, augmented reality, cyber security, and advanced manufacturing (Ferreira et al., 2023; Psarommatis et al., 2023).

2.2 HISTORY AND EVOLUTION OF THE INDUSTRIAL REVOLUTIONS

Landes (1969) believed that since the fire of knowledge had been stolen from Mount Olympus (the dwelling place of ancient Greek gods) right from their eyes by Prometheus, he bestowed it upon humanity. Man interacting with this began to innovate and renovate using this knowledge to better their environment and nature. With this, humanity had no choice but to keep innovating. Talking of the Industrial Revolution is naturally reasoned as a cogent break in the history of humanity, basically from the Neolithic period (Weightman, 2010).

2.2.1 First Industrial Revolution

Most English citizens were living in poverty before the Industrial Revolution. Changes in society, culture, and government occurred during the Industrial Revolution. The new epoch is an Industrial Revolution invention, in the course of which a wide variety of businesses flourished, including Henry Cort's puddling and rolling technique for producing iron (1740–1800), Samuel Crompton's mule for spinning cotton (1753–1827), and James Watt's steam engine (Mohajan, 2019). According to Patrick O'Brien (O'Brien, 2017), the Industrial Revolution in Britain can be traced back to the country's prosperous and adaptable agriculture and its plentiful and easily obtainable resources of coal and various other minerals, international trade, and significant technological discovery and innovation. The advancement cut across some manufacturing areas, such as mineral extraction, whereby available mineral resources are taken out for processing. It also affected the fibre industry, in which textiles were made in large quantities, even in people's homes. During the course of the revolutions, there was an increase in work hours, which was not much and was hectic, and workers had to work 16 hours to save their families and to survive. These revolutions also decreased the mortality of childbearing in which women gave birth, and their children lived because strong babies were the ones living after being birthed during those years of the Dark Age. Another implication brought to humanity through these revolutions was decreased death rates (Lehmann-Hasemeyer et al., 2023).

By the turn of 1840, the tidings spread to the United States, some European mainland countries such as France and Germany, and also touched some parts of the Asian continent (Japan precisely) later extended on to China by the turn of the 20th century, this advancement had travelled wide into some other continents, bringing about modifications to human's lives and activities and making them more effective and productive in their businesses and also in their interactions with nature and themselves. With this advancement, other countries have also seen the development of new technologies around the globe. The Industrial Revolution spans four generations, making each generation contemporary. On investigation, it was found that we have entered the fifth generation. The stages through which the Industrial Revolution spans consist of the First Industrial Revolution (Industry 1.0), the Second Industrial Revolution (Industry 2.0), the Third Industrial Revolution (Industry 3.0), and the Fourth Industrial Revolution (Industry 4.0), which, according to research, is at the entrance of Industry 5.0 (Viollet, 2017).

The initial date of the First Industrial Revolution has been debated among many researchers, historians, and scholars, but many confirmed that it began in 1760 and ended in 1840. The First Industrial Revolution, which began in Britain in the 18th century, started with steam machines powered by water as a means of movement. It was said to have been eight times better than manual labour. Helping in the manufacturing industry during this period means travelling a longer distance in fewer hours, making manufacturing processes easier and faster. The era was made to be one of transformation from an artisan (hand means) into a machine means of production. During this period, many industries sprung up. As a result of this innovation and the invention of machines that produce goods in large quantities, the production sectors have been made efficient and productive. This first revolution was a total turnaround for humanity, making mankind's society turn into a modernised society and, as such, making life easier for mankind.

Almost all the sectors caught the widespread fire of the revolution; even the GDP received an increment, which also caught the population's attention. In their studies, Agarwal & Agarwal (2017) assessed firms' technological change and development during the First and Second Industrial Revolutions. They stressed that the invention of the power loom and steam engine and improvements in iron-making technology were the primary reasons behind the First Industrial Revolution. The revolution comprised innovators, skilled artisans, bankers, merchants, and wealthy individuals who owned the new factories and mines. The growth of England may be traced back to local investments in infrastructure, including roads, bridges, canals, port works, and eventually railways (Pearson & Richardson, 2001). The revolution relied on Newtonian science, widely disseminated exclusively in Britain. The United States and other countries lacked the necessary scientific background to launch the revolution without it. Britain also provided superior protection for intellectual property and real estate investments (Beker & Lipsey, 2002). Waterpower and coal for fuelling the new machines; iron ore for manufacturing machines, tools, and structures; rivers for interior transit; and harbours from which commercial ships set sail were only some of England's abundant natural resources in the 1700s. According to the work of Clark et al. (2008), England was the most politically stable country on the globe and the dominant colonial power. The colonies under its control might provide it with raw resources and operate as a distribution hub for produced commodities.

Additionally, England already had a developing economy that could support industrialisation. Entrepreneurs poured much money into researching and developing cutting-edge industrial innovations. In England, there was a rising enthusiasm for scientific inquiry and innovation (Montagna, 2013). The financial system in England was quite advanced, which allowed for substantial industrial investment. The opportunity to borrow money from banks spurred the purchase of new equipment and the development of new enterprises. Those who ran successful factories reaped enormous rewards. They saved a tonne of money by stocking up on machinery for the plant and buying materials in bulk. They had put money into creating international commerce and infrastructure, including canals, railways, and steamships. They became a new economic elite in Britain (Montagna, 2013).

Deane and Cole (1962) establish that Britain's political stability was a key factor in the country's capacity to sustain a high economic growth rate. British transport was drastically altered with the introduction of the locomotive and the development of railroads. Many new jobs were generated to train employees and miners, and improvements were made to the agricultural and fishing sectors due to the railroads. Due to its convenience and comfort, rail travel gained popularity. The revolution's focus on textile development was key to its success. Britain's textile industry has helped the country's economy expand in recent years (Sugden & Cockerill, 2017). During the Industrial Revolution, iron and steel production increased considerably. Iron output in Britain climbed from 28,000 metric tonnes in 1750 to 250,000 metric tonnes in 1805. Sulphuric acid, hydrochloric acid, alkali, sodium carbonate, sodium sulphate, potash, bleaching powder, concrete, etc. were mass-produced throughout the revolution. These substances produced glass, textiles, soap, bleaching fabric, etc. John Roebuck (1718–1794), an English inventor and manufacturer, developed the lead chamber procedure for making sulphuric acid in 1746. Sodium carbonate was first manufactured in 1791 by French scientist and physician Nicolas Leblanc (1742–1806). Iron and steel were "pickled" in sulphuric acid, and the acid was further used to bleach textiles. Since all industries advanced throughout industrial revolution (IR), it was crucial that transportation also advance so that people, animals, and things could be moved about more easily (Agarwal & Agarwal, 2017). Advancements in economics, culture, society, and industry cannot occur without adequate transportation (Choudhary & Rao, 2018). To conclude, during the First Industrial Revolution (1760–1840), Great Britain experienced economic growth and prosperity, leading to the modern era. The development of industries like textiles, iron and steel, chemicals, agriculture, and transportation improved living standards. However, most populations were poor, and child and adult mortality rates decreased. The global population increased rapidly, and a middle class emerged. Despite its successes, the Industrial Revolution had negative effects, including increased unskilled labour, women's and child labour, the slave trade, and environmental pollution. Trade unions were formed to fight worker torture, and capitalism emerged for wealthy individuals, while socialism was created to support poor labourers.

2.2.2 Second Industrial Revolution

The economic slump began in the latter half of the 1830s and continued through the early 1840s. While the stock market fell, it gave birth to the Second Industrial Revolution. The starting point of the period in question was signalled by a surge in the pace at which the number of existing industrial machinery grew. Developing electrical energy and new manufacturing techniques during this period also increased output and improved quality. Because of the time of industrialisation's standardisation, this era is also often called the "Age of Technology". The second phase of the Industrial Revolution saw a rise in specific industries and enormous gains in scale and productivity. To borrow the well-known phrase used by Alfred Chandler preceding the outbreak of World War I, research dated this period from 1870 to 1914. Over time, the primary impact of the developments above involved communication and electrical energy for running the initially developed machines, which referred to the creation of new machines and the renovation of the existing ones. Given technological improvements and the capacity for large-scale production, we witnessed the need to utilise specific renewable resources such as gas, water supply, etc. All these advancements made it possible for us to work more quickly. Manufacturing economies of scale are becoming more significant and are to blame for this trend. Some of them were essentially physical, such as in the building of vessels and cylinders for chemicals. For example, the price is related to the outer area, whereas efficiency equates to volume. People were supposed to follow employment, and in the early 1900s, workers were seen moving from homes in the countryside to metropolitan regions in the quest of manufacturing positions. Compared to only 6% in 1800, around 40% of Americans resided in cities by 1900.

Despite the emergence of huge enterprises like Carnegie Steel, DuPont, Ford Motors, and General Electric in the United States and the corresponding European firms, it should be emphasised that only a small portion of the workforce was employed by these companies by 1914. The typical company in the industrialised West remained restricted, a niche player, frequently specialised yet flexible, and competitive, and the period experienced a wide range of fluctuations in the economy, including the "Great Depression" of 1893 and the market "Crash" of 1930, not to mention both World Wars. Capital has played a crucial role in this transformation, and competition has increased, contributing to globalisation. Since the middle of the 19th century, industrialisation in Europe and the United States has accelerated, and the number of industries has increased (Hobsbawm, 2016; Klingenberg & Antunes, 2017). People's standard of living largely depends on food availability and proper nutrition. In the 19th century, emerging technology impacted food supplies in manufacturing, supply, safeguarding, and preparation. The textile industry also received its incentives in a new way. After 1870, advancements in the textile industry needed to be made faster and distinguished by breakthroughs. The sewing machine was a significant development at this time. Despite a global quest for a machine that would replace the motion of the human hand in the stitching process, clothing production was falling off the rest of the textile sector during the early phases of the Industrial Revolution. Technically speaking, these machines were first unusable, but around 1830, a solution started to emerge (Klingenberg & Antunes, 2017).

The Second Industrial Revolution was, in many ways, a continuation of the First. Comparable similarities could be seen in many industries. Firstly, this period saw a significant difference in wages and quality of life from 1870 to 1914. Secondly, the Western industrialised countries kept their dominant positions in terms of technical leadership; although the geographic focus was no longer exclusively on the United Kingdom, it had become more dispersed. Lastly, it forever changed the relationship between natural knowledge and technical practices, leading to several other Industrial Revolutions in the following years.

2.2.3 Third Industrial Revolution

The Third Industrial Revolution came into being in the 1950s, only a few years after the end of World War II, and it was partly automated using memory-programmable controllers and computers. As a result of such advancements, it is now possible to automate the production process entirely so that only a minimal amount of engagement from human labourers is required. The introduction of computers marked the beginning of the Third Industrial Revolution, often referred to by the term Robotics Age (George & George, 2020). During the latter phases of the second wave of industrialisation, technology has transformed conventional systems into computerised ones, enabling these systems to work independently. Based on the premise that the introduction of new forms of communication tools corresponds with the introduction of new energy regimes, most notably renewable energy sources. Various electronic devices were produced during the period known as the Third Industrial Revolution, including semiconductor circuits, computer logic systems, Metal–Oxide–Semiconductor (MOS) transistors, and the connection to the Internet. These devices comprise computers, digital cellular phones, and microprocessors. Internet connection was also established during this period. To put it another way, the advent of the computerised era brought about the transition from the analogue period into the contemporary cyber world (Myers et al., 2000).

Situations arose, even to the point of totally substituting human actors. In the 1960s, the Programmable Logic Controller emerged. Because computer hardware is so widely used in production systems, matching software had to be developed. This factor led to the industry's digital information and technologically driven industrialisation (Sharma & Singh, 2020).

2.2.4 Fourth Industrial Revolution

The advent of the Fourth Industrial Revolution ushered in several waves of societal, governmental, cultural, and economic disruptions throughout the 21st century. The

Fourth Industrial Revolution will be primarily driven by integrating computerised, biological, and physical innovations, building on the ubiquitous availability of digital technologies brought about by the previous industrial or digital renaissance (Adeniran, 2016).

Similar to the steam-powered manufacturing plants of the early Industrial Revolution, the incorporation of scientific knowledge to large-scale production and manufacturing during the period known as the Second Industrial Revolution, along with the beginning of digitisation during the third wave of industrialisation, the Fourth Industrial Revolution witnessed the emergence of innovations that involve AI, genome modification, augmented and virtual reality, robotics, and 3D printing, which are rapidly altering how humanity creates, exchanges, and distributes value. As with previous revolutions, this will profoundly affect organisations, industries, and individuals. Moreover, this revolution will be shaped by the decisions people make today: the globe 50–100 years from now will contribute much of its distinctive nature due to how people feel about, invest in, and implement these potent new technologies (Nuvolari, 2019).

It is crucial to recognise that the Fourth Industrial Revolution entails a systemic shift throughout numerous industries and facets of everyday life: the crosscutting effects of emergent technologies are even more significant than the thrilling capabilities they represent. The capacity of humanity to modify the fundamental elements of life has lately been greatly enhanced by inexpensive genetic sequencing and methods that include Clustered Regularly Interspaced Short Palindromic Repeats (gene editing tool). Intelligent technology enhances methods and competence throughout each industry. Neuroscience has made uncommon strides in employing and impacting the neurological system as a component governing human biological research. Computerisation is altering old-century moves and factory paradigms. Technological advances with the value of blockchain technology, used in executive compensation, are reshaping the world (Nuvolari, 2019).

The global shift in society results from a combination of various factors. Its impact on the incentives, regulations, and norms that shape our financial existence alters our patterns of interaction, acquisition of knowledge, entertainment, social connections, and self-perception as individuals. Furthermore, the notion that innovative ideas are being generated and employed increasingly rapidly has implications for individual viewpoints, communities, and political systems. Hence, the interconnection between our responsibilities towards each other, our potential for personal growth, and our ability to positively impact society are closely intertwined with and impacted by our involvement with the advancements associated with the Fourth Industrial Revolution. The current upheaval is neither solely occurring in our midst nor are we passive recipients of its effects. On the contrary, we are afforded the opportunity and obligation to furnish organisation and guidance (Sharma & Singh, 2020). However, some analysts have noted that this revolution could increase inequality, especially its potential to disrupt labour markets. As automation replaces labour throughout the economy, displacing employees by machines may exacerbate the disparity between returns on capital and returns on labour. On the contrary, it is conceivable that removing employees by technology could result in an overall rise in secure and gratifying employment (Groumpos, 2021).

All preceding industrial revolutions have positively and negatively affected various parties. The inability to allocate the resulting benefits equitably or predict repercussions has culminated in global challenges. By recognising the risks, whether cybersecurity hazards, misunderstandings on an enormous scope via the internet, potential unemployment, or rising societal and economic inequalities, we can take the necessary steps to guarantee that the Fourth Industrial Revolution primarily benefits humans.

Therefore, the Fourth Industrial Revolution is not a forecast for the future but a call to action. It is a vision for developing, disseminating, and governing technologies to cultivate an increasingly strengthening, working together, and durable basis for the growth of society and the economy, based on the shared principles of communal well-being, human dignity, and lifelong stewardship. Understanding this aspiration will be the greatest, excellent, and most significant challenge and obligation of the next 50 years (Sharma & Singh, 2020).

2.3 INDUSTRY 4.0 FEATURES

The underlying idea to make industries dynamic, more adaptable, and smarter is Industry 4.0. According to Wang et al. (2010), Industry 4.0 is an emerging phase of industry in which manufacturing operations systems and information and communication technologies, mainly the IoT, are integrated and referred to as CPS. For integrating cutting-edge features into the CPS, the Web of Things was utilised, which facilitates operations of physical reality connection with computing and communication infrastructures (Pandey et al., 2023). An increase in industry connectivity requires 4.0 technologies to safeguard vital industrial systems and production processes against cyber-attacks. The aim of Industry 4.0 is to make "smart factories" that are more adaptable, productive, and sustainable. Its objectives involve simplifying production procedures, maximising resource utilisation, raising product quality, cutting costs, and boosting industries' competitiveness. Industry 4.0 can also potentially change business structures, logistics, and supply chains, with profound societal and economic effects. Industry 4.0 key attributes are as follows.

2.3.1 Cyber-Physical Systems

Cyberspace is the widely utilised, networked, and computerised digital technology that distinguishes infrastructure for computing and communication. A few industries that employ CPS are manufacturing, energy, infrastructure, consumer products, smart building, military, robotics, communications, healthcare, and transportation (Karnouskos et al., 2021). A CPS emphasises networking many components, contrary to traditional embedded systems created as standalone devices (Lee, 2008). In other words, it is an embedded system with network connectivity that can transmit and receive data. The "Internet of Things" is a term that is frequently used to describe the CPS's connection to the Internet.

2.3.2 Internet of Things

The IoT is a collection of technological devices with an information centre that monitors the state of physical items, collects valuable data, and transmits that data to software applications across IP networks. The IoT quickly overtakes all other computer platforms. IoT technologies drastically alter our way of life with recently created applications, including Smart Transportation, Smart Cities, Smart Houses, and Smart Grid. The ubiquitous connectivity of physically dispersed smart IoT objects expands computing and communication capabilities to IoT items with different specifications. These devices' sensing abilities enable the direct or remote collection of real-time data from the physical environment. By analysing the data gathered, we can create a more intelligent environment and govern it more effectively.

2.3.3 Artificial Intelligence

AI is at the heart of Industry 4.0, combining a wide range of tools that let computers mimic human intelligence and learn how to do specific tasks. This technology has the potential to increase productivity in the manufacturing sector. Industry 4.0's continual technological advancements significantly impact the expanding manufacturing industry. AI is one kind of cutting-edge tech that has the potential to improve productivity, product quality, and business finances significantly. Several machines are constantly in dialogue in a smart factory with hyperconnected manufacturing processes. To improve quality control, standardisation, and maintenance, manufacturers are undergoing a digital transformation that involves managing and utilising their data sets via AI and machine learning (ML).

Many services in Industry 4.0, such as those provided to customers, stand to gain from incorporating AI into production procedures shortly (Zhang & Lu, 2021). It makes our work quicker by allowing us to get better outcomes with fewer people. Because it relies on digital technology, Industry 4.0 is more intelligent and productive. As AI develops, it enables computers with enhanced senses, including vision, hearing, learning, and the ability to access new and exciting learning environments. Connected industries that are deeply integrated with the supply chain, design team, manufacturing line, and quality control into an intelligent engine that gives practical insights with the assistance of AI are a must for Industry 4.0.

Manufacturers require collaboration throughout the supply chain cycle to take advantage of the prospects presented by Industry 4.0 fully. Asset management, supply chain management, and resource management are now the primary applications of AI, ML, and IoT. Improved asset monitoring, supply chain transparency, and inventory utilisation are all possible with the help of today's state-of-the-art systems. Algorithms, AI-driven procedures, and quality optimisation are examples of how ML may enhance predictive maintenance (Javaid et al., 2022). AI makes it possible to monitor factory floor operating loads swiftly and accurately in real time, improving production planning efficiency. Producers may boost output, PM, and asset burdens by integrating ML with total equipment effectiveness.

2.3.4 Cloud Computing

Cloud computing is helpful for Industry 4.0 firms that use collaborative supply chains because it provides near-real-time access to data, allowing for proactive management and rapid adaptability to changing conditions. Sixty per cent of factory executives throughout the globe say cloud infrastructure is vital for realising Industry 4.0's full potential, and it is one of the top five technologies with an acceptance rate of 85% or higher. Increasingly crucial in manufacturing lines and driverless vehicles, robotics and ML rely heavily on the computational infrastructure provided by the cloud.

The IoT is transforming logistics and shipping by connecting machines directly to one another instead of humans. In the era of Industry 4.0, producers and merchants can monitor individual items through the supply chain using technologies like location tracking and cloud computing infrastructure. Cloud-based fleet management solutions simplify handling difficulties and adjusting to unforeseen circumstances by providing real-time data on vehicle availability, service historical events, maintenance documentation, and mileage.

2.4 FUNCTIONS AND BENEFITS OF INDUSTRY 4.0

2.4.1 Functions of Industry 4.0

It is believed and generally said that the functions and advantages of the industrial revolutions, primarily Industry 4.0, are geared towards making the world easier and better for people to live in. The Fourth Industrial Revolution, also known as the "digitisation" era, has its embodiment in what is called the "cyber-physical systems" which would be named AI, which is referred to as a programme that enables the machine or device to perform as a human being. It operates like the human brain is its power source, making it subhuman in computers (Kurt, 2019; Mavropoulos & Nilsen, 2020). All these developments are for purposes it has already been serving in every sector of life, especially in food supply sectors (agricultural sectors), engineering, education, marketing, and even the military, whose functions and impacts will be discussed in this chapter.

Food production, generally known as "agriculture", has had a more significant role in the world ever since the creation of the universe (Corallo et al., 2018; Bag et al., 2021). No country or nation can live beyond its GDP. This is the yardstick used to measure the development and progress of a country. Nevertheless, agriculture, the nation's primary food source, needs a more advanced, faster, and more efficient way of production. With Industry 4.0, food production could be carried out with higher quality, resulting in the mass production of food for the whole country. Weather or climate is one of the significant challenges facing agriculture. Industry 4.0 could help get climate changes that would help them not plant at the wrong time, with the weather forecast given by Industry 4.0, thereby bringing about profitability for the farmer (Corallo et al., 2018; Bag et al., 2021).

Industry 4.0 could help crop and soil monitoring because of the daily soil degradation. With the innovation of Industry 4.0 regarding agriculture, farmers can detect any dearth in the soil, thereby aiding the productivity of the soil by adding fertilisers to the soil. There is a place for disease control by using Industry 4.0 in agriculture. With the aid of the prediction of AI, farmers can spot the disease on the farmland and avoid those places when planting crops, which would also help the farmer do justice to the affected areas. Industry 4.0, as discussed previously, has a more significant role to play in the production of food in the world today and, if utilised well, would go a long way in increasing the productivity of the nations of the world, making it a better and more ideal place to live (Adeniran. 2016; Kurt, 2019).

Manufacturing is an essential component of modern existence. Manufacturing may encompass textiles, automobiles, electronics, and other sectors. All of these are fabricated and cannot be created by human hands. Industry 4.0, once implemented in this domain, can significantly enhance the delivery of quality materials. For example, in the automotive sector, machinery is employed in manufacturing. Nonetheless, with the assistance of Industry 4.0, automobiles can be produced with minimal complexity and reduced human labour, as Industry 4.0 integrates machinery and AI. Rather than having humans transport various components of the automobile, robotics programmed for this task are utilised to elevate the materials to the optimal positions, resulting in expedited and more automated movement (Ferreira et al., 2023; Javaid et al., 2022; Bigliardi et al., 2023; Pandey et al., 2023; Psarommatis et al., 2023).

Industry 4.0 could also detect any fault in the production processes that give the product the best chance of being manufactured and bring out the best quality, which could prompt good feedback from the populace or consumers to the company. The textile industry, which began with the First Industrial Revolution but spanned throughout all the industrial revolutions, also received technological advancement whereby the operator does not have to sit by the machine; all that is needed is to enter the required design, and it continues its work with less human effort; this could reflect in the engineering world, helping in every aspect of engineering (Ferreira et al., 2023; Bigliardi et al., 2023).

Civil engineering, automobile engineering, electrical engineering, and various other disciplines can detect errors or flaws and assist many in determining precisely what is wrong, thereby facilitating better decision-making. It also comes with designs for the civil engineer, their AI combined with big engines which the engineer just has to monitor, and it can even work independently without being affected by anything (which many labourers have to do) at a very high speed. If it can be incorporated into computer engineering, it could aid fast decision-making based on the problem it poses to solve. It has extraordinary abilities and functions in this facet of life, bringing about new and innovative technologies.

Industry 4.0 has gone a long way in aiding improvement in the health sectors in terms of diagnostic function, taking patient records, moving upward to treating patients, and prescribing methods to be used to treat patients. It has begun aiding treatment personally at home without having to go to the hospital for treatment, only after a special consultation with the doctor. Merative is a tool under Industry 4.0 used in today's health sector that is needed for human monitoring and analysing patient data. It could analyse large patient data because Industry 4.0 combines AI and machines. Less human effort is required to control it. Enlitic is another tool in this contemporary age used in diagnosing and giving accurate information about the patient's status. Viz.ai is another tool based on cloud data, saving patients' data for record and reference purposes. Industry 4.0 also plays a role in the health phase of life, bringing effectiveness to the health areas and helping save lives (Bote-Curiel et al., 2019; Roy et al., 2022).

Education, in this regard, also received improvement with Industry 4.0. Based on the works performed by Industry 4.0, one of them is computing extensive data and analysing them. This applies to students' data and performance records. There is a place of self-learning in education today whereby students can be in their comfort zone learning while interacting with AI. Also, there is room for instant response, generally known as "feedback", when the AI tool gives a test, which is one of the tasks performed by Industry 4.0. With Industry 4.0 in the education world, special needs children have access to education even without visiting the four walls of a school. It also gives students access to education 24 hours a day, wherever they may be in different parts of the world. With Industry 4.0, students can also reach out to the world, interact with people worldwide, and learn new things such as languages and technologies. Also, tutors can get more information about what they teach the student. It even helps students be accurate in their tests, making them error-free (Pandey et al., 2023; Psarommatis et al., 2023). The significance of creativity in education, facilitated by Industry 4.0, cannot be overstated, as it fosters engaging and enjoyable learning experiences for students. Nonetheless, Industry 4.0 is unreliable in this domain of existence. This is due to a deterioration in students' performance and their capacity to articulate their knowledge across various disciplines.

Marketing is another facet of life in the business world. Industry 4.0 has had a significant impact; not just a few businesses are incorporating technology in their activities. "Without marketing, it is said that no business can go far". It would need an advertisement for people to know about the manufactured product. There is a need for Industry 4.0 to create a good and error-free awareness for the populace to be attracted to a product. Industry 4.0 in marketing can improve and produce targets for personalising market campaigns. By the geographical elements, preference, and purchase records, Industry 4.0, if incorporated into marketing, can analyse vast customer data. In making marketing efforts fruitful, implementing Industry 4.0 can help businesses run their marketing content and connect it with particular, specific consumers, which will bring about effectiveness in the marketing techniques. There is also room for plagiarism checks using Industry 4.0 so as not to copy other people's words to attract customers to their products. In this regard, Industry 4.0, in the marketing or business world, helps the rates at which conversion on the product marketed can grow, and not just conversion rates alone, but also upward movement on consumer tastes of the product will increase (Ernst & Frische, 2015).

Industry 4.0, which cuts across every aspect of human life, does not leave behind transportation, starting from automobiles created in this contemporary age. Some self-driving vehicles drive themselves with or without human control and will have to park themselves even when they reach their destination. Industry 4.0 also helps reduce accident rates, including pedestrian-vehicle and vehicle-vehicle collisions. With the implementation of Industry 4.0, the vehicle can track a pedestrian even when the driver overlooks the event about to unfold. However, with the help of Industry 4.0, both pedestrians and drivers are safe from accidents.

In road management, Industry 4.0 is implemented for the safety of both pedestrians and car users with the AI-controlled traffic light. If any emergency occurred due to the sensor, it would notify if anybody was crossing the road. This system could also help avoid vehicle collisions and other road accidents for road users. Industry 4.0 incorporated in the aviation sector can also help predict delay times in making a route that could bring about delays on the road and air. With its analysis based on the map data, it gives suggestions for other routes to avoid delays that may occur. Industry 4.0 can help get the original license. The number of people with licenses to drive is significant, and humans can wear out trying to check the authenticity of a license. However, Industry 4.0 can go as long as possible and even faster than one might think. It could assist in road monitoring for road users and even the government to ensure proper knowledge of the roads to be attended to (Ceruti et al., 2019).

The military sector in the world has also been changing due to technological advancements. With the look of things, it is one of the significant areas in which technological advancement has ravaged, with the use of drones, autonomous armour vehicles, automated guns, etc. It even goes a long way to ease the men of the work of going to the field fighting with the drones at work. Surveillance of the enemy's territory and whereabouts and checking the areas where mines are planted to avoid many casualties or deaths at the war front is now possible.

Industry 4.0 functions as the best decision-maker in this area of life. There is much data to process, even at the war fronts when aiming at a target. Industry 4.0 comes with a device for the automatic target, which could help at shooting aright. Industry 4.0 can analyse multiple scenarios and even risk-taking decisions on this side of the world. Industry 4.0 also functions as a training implemented for military personnel, training them to work in critical situations and make decisions. There are virtual training tools or simulations that could also help in combatant actions. Since Industry 4.0 covers all aspects of life, cybersecurity is also an addendum in the military in our contemporary age. Industry 4.0 is also tasked with shielding and defending data and confidentiality in every military section.

Government is one of the important aspects of life that Industry 4.0 positively affects, and it functions as a helpful tool in analysis and decision-making. Industry 4.0 in this sector can take data of dwellers in such a country or state and help the government to appropriately know about amenities that should be provided for a particular area under their control. Concerning the fact that the government covers almost every aspect of life, with Industry 4.0, the government can investigate every other sector under it and give it what it needs. Industry 4.0 is also effective in the monitoring of projects and properties. There can be properties that need to be charged for their payment of taxes but are not shown to the world. Industry 4.0 can detect such properties to avoid people evading taxes, and it could assist the government in making policies. Since the government deals with many people, manual human labour cannot get their data. It could pose negligence because they have no record of the people to access, let alone know their needs. However, with Industry 4.0, they can access extensive data of people and even know what they need in their areas, respectively, and assist those processing the data.

Industry 4.0 also assists the government in policymaking; thereby, having the capacity to get to the societal trends could help the government make efficient and productive policies in favour of the people. With Industry 4.0 being incorporated in the governmental lines, it would help save hours of work, make quick decisions, and process data that are very large for a man to process. Banking sectors cannot be overlooked in stating the areas of life affected by Industry 4.0. Industry 4.0 functions as a transaction record taker and management investment, which could help banks and investors know the trend and engage them with the decision to invest in a particular period. Even the insurance companies are not exceptional in what Industry 4.0 is doing.

Industry 4.0 also helps in cybersecurity and detecting fraudulent activities, saving the banks from the risk of bankruptcy. A customer care service that works 24 hours daily without interruption and human control also exists. It necessitates chatting with customers and satisfying the customer based on whatever complaint. This is popularly known as "Chatbot". This saves people from working longer hours and could give them space to face something else. Industry 4.0 cannot be overlooked in the loan and credit sectors of the bank, in which it checks customers' loan histories to help the bank decide on giving out the loan (i.e., if the customer is loan-worthy) based on the previous transaction. Implementing Industry 4.0 also aids customers' experiences with withdrawing and depositing their money while no one attends to them instead of being delayed in the queue for a long time. Banks can regulate their compliance, which also helps the government fish out those using the bank to execute fraudulent activities and prosecute them.

Water supply is not left behind in implementing Industry 4.0 in the contemporary era. As said in every other aspect of life, incorporating Industry 4.0 also takes the form of a decision-maker knowing where water is needed and predictive measures for those in charge of this management.

Exposition alongside prognosis instruments to determine precisely what occurred in a particular place at a specific moment in the water supply system in the absence of instruments and monitor data is possible. Also, there are prediction instruments to simulate a possible strategy and execution model, the one that powers the water supply network and prescriptive tools for decision-assisting channels, which are gaining popularity by offering the most suitable solutions available to meet a specific problem or limitation and even, in some cases, automating a choice. These tools help people make better decisions, and these are stated to be the functions and advantages of Industry 4.0 to water supply.

2.4.1.1 Benefits of Industry 4.0

Interoperability is a key goal of Industry 4.0, which seeks to advance the concept of interconnection across the numerous parts and devices found in factories. This allows devices, infrastructure, and people to communicate and work together without interruptions throughout the value chain. Industry 4.0 makes it possible to predict upkeep using the IoT and data processing in various contexts. Machines and other equipment may be tracked continually, and the possibility of faults or needs for maintenance can be identified in advance. This makes it possible to perform repairs or replacements promptly. This strategy assists in bringing down service costs while also cutting down on disruption.

The Fourth Industrial Revolution technologies provide more effective resource administration and energy optimisation. Data collection and analysis allow firms to pinpoint regions with garbage, improve the efficiency of their operations, and cut their overall energy usage, which results in savings in expenses and advantages for the environment. Industry 4.0 makes flexible and versatile production procedures possible. Companies can swiftly launch new goods, customise orders, and effectively manage smaller quantities when they have the flexibility to quickly change processes, rearrange machinery, and react to fluctuating market needs. Industry 4.0 can bring about a change in the conventional structure of labour. Though certain mundane activities could be digitised in the future, human employees will primarily concentrate on activities that require inventiveness, the ability to solve issues, and analytical ability. To adapt to the shifting requirements imposed by the digital era, workers must acquire new skills and improve their existing ones.

Improvements in the overall output and quality of the product may be made with the help of Industry 4.0's continuous data collection, analysis, and process optimisation. Intelligent systems can recognise errors or variances promptly in the manufacturing process, which helps minimise waste and ensure that quality requirements are maintained consistently. The technologies of Industry 4.0 make it possible to increase the supply chain's availability and synchronisation. Actual communication of data and analytics improves performance, reduces delay periods, and makes it possible to promptly manufacture and transport at every stage of the supply chain, from procuring raw materials through production, transportation, and distribution. The advent of Industry 4.0 paves the way for the development of brand-new companies and additional income sources. By utilising interaction along with data to supply services that add value and sustain long-term connections with customers, companies, for instance, can move away from supplying things to delivering product-as-a-service or subscription-based models.

Industry 4.0 provides predictive maintenance by using analytical tools and detectors in its machines. This feature is known as predictive maintenance. It is possible to anticipate prospective machine breakdowns or maintenance requirements if the machine's performance is continually monitored and the data patterns are analysed. This strategy reduces the number of unscheduled delays, cuts down on maintenance expenses, and maximises the use of assets. Industry 4.0 will improve the handling of supply chains by allowing real-time monitoring and tracking of all commodities across the supply chain. This will allow for more effective overall supervision of the supply chain. Companies can improve their stock administration, demand prediction, logistics, and delivery operations by integrating IoT, RFID tracking, and sophisticated analytics. Consequently, this increases productivity and decreases lead times, resulting in higher customer satisfaction.

Industry 4.0 makes it possible to provide a more individualised and streamlined experience for customers. Firms can obtain insights into their customers' tastes, behaviours, and demands via data analytics, which enables them to provide more personalised products, engage in more focused marketing, and provide enhanced customer service. This results in increased client satisfaction, loyalty, and subsequent business. Firms may speed up their approach to getting noticed for new goods using the technology of Industry 4.0. Rapid prototype development, adaptive design changes, and a speedier production increase are all possible, thanks to combining additive production with digital modelling and design technologies. Because of their swiftness and adaptability, businesses can react rapidly to changes in the market and acquire an edge over rivals as a result.

Industry 4.0 strongly emphasises security and risk control in manufacturing environments. This will result in improved safety and management. The combination of IoT indicators, statistical analysis, and ML enables early detection of possible safety hazards and risks, which enables proactive mitigation of such hazards and risks. This contributes to the creation of safer working conditions, the reduction of accidents, and an overall improvement in the health and safety of the workplace. Industry 4.0 allows businesses, vendor partners, and customers to collaborate and form partnerships. The interconnection of everything and the real-time exchange of data make seamless cooperation possible in areas like product creation, coordination of supply chains, and innovation of services. The result of this partnership is an increase in overall ecosystem efficiency, innovation, and value production.

Industry 4.0 allows small and medium-sized enterprises (SMEs) to compete globally, empowering these businesses. SMEs can now implement Industry 4.0 practices because digital technologies, cloud computing, and data analytics are available. These practices allow SMEs to optimise their operations, increase productivity, and enter new markets, levelling the playing field with more prominent organisations. Industry 4.0 promotes sustainable manufacturing practices by optimising energy usage, decreasing waste, and enhancing resource efficiency. These three goals may be accomplished through the reduction of waste. Actual tracking, statistical analysis, and automation make it possible for businesses to pinpoint energy-intensive activities, put in place energy-saving measures, and reduce their negative influence on the environment. This helps comply with environmental rules and contributes to achieving sustainability objectives.

Industry 4.0 may help businesses and industries become more competitive globally by adopting and using technology. Businesses can boost their production, cost-efficiency, and quality levels by using cutting-edge technology, improving existing procedures, and embracing innovation. This strengthens their position in the worldwide market, paving the way for increasing market share, expanding their firm, and continued achievement over the long term.

The benefits of Industry 4.0 may be applied to a wide range of facets of corporate operations, including manufacturing and the management of supply chains, as well as consumer involvement and environmental responsibility. Companies have chances to promote innovation, increase efficiency, and adapt to the changing digital world due to the revolutionary nature of Industry 4.0. It is essential to remember that the execution of Industry 4.0 could be carried out differently by different organisations and sectors, according to the requirements of each and their level of preparation. Despite this, the fundamental ideas of connection, digitisation, and automation will continue to play an essential role in the paradigm shift towards more automated industrial practices.

2.5 APPLICATIONS OF INDUSTRY 4.0 IN ENGINEERING

There have been numerous applications of Industry 4.0, ranging from its application in agriculture, tagged “Agriculture 4.0”, to its application in the supply chain and a host of others. This section investigates the several applications of Industry 4.0 in engineering, involving production and issues around it.

2.5.1 Case Study of Smart Scheduling

Rossit et al. (2019) did insightful work on “Smart Scheduling”. Smart Scheduling merges the principles and instruments of Smart Manufacturing alongside more conventional scheduling practices. Smart Scheduling uses both the adaptability of Cyber-Physical Production Systems (CPPS) and the speed with which Decision Support Systems (DSS) can solve dynamic scheduling issues. The constant two-way dialogue between production floors and their respective DSS amplifies background noise or causes the system to crash. However, Smart Scheduling can fix this issue and rewrite schedules via Inverse Optimisation and Inverse Scheduling.

Industry 4.0 heavily depends on CPS, which link digital and physical elements. Over IoT, CPS may keep an eye on the workings of the actual world, simulate it, and make choices without any central authority. These CPS may produce simulated versions of the production procedure, allowing for easy remote management using the Cloud. This enables us to conceptualise the production set as "Cloud Manufacturing in Action". There will be shifts in production schedules due to the introduction of these novel manufacturing models. Production planning problems are notoriously complex, but Rossit et al. (2019) proposed a clever method for addressing them, even in tiny instances such as Flow Shop scenarios with no fewer than four machines. The long-term success of businesses in highly competitive marketplaces depends on the reliability of the remedies they provide. The integration of scheduling procedures with the smart production procedures of Industry 4.0 highlights the paramount importance of such procedures, which are activated using Manufacturing Scheduling Systems (MSS).

Human schedulers skilfully need the computer to make different decisions in MSS. The combination of MSS and scheduler can handle any static or dynamic scheduling challenge. When CPPS are combined with the Tolerance Scheduling issue, a fundamental shift occurs in product production. Choosing an operational timetable is a significant part of the planning process. However, it is also one of the most challenging parts because of the sheer amount of data to consider. Industry 4.0's dynamic scheduling is another important component since it requires preparation in the face of unpredictability. This is not a minor issue that, if left unchecked, might result in costly consequences. Job availability, resource limits, and production schedule shifts are just a few aspects of the manufacturing process. These authors propose two formidable methods of scheduling. These include Reactive Scheduling and Predictive-Reactive Scheduling. Reactive Scheduling is a method that makes choices as events emerge, and Predictive-Reactive Scheduling is often used in industrial systems. In Predictive-Reactive Scheduling, schedules go through several iterations, adjusting based on what happens. In the first phase, a timetable is developed as a solution without considering potential delays. The second phase, "reactive", involves adjusting the initial timetable according to unanticipated circumstances. This procedure is carried out every time a new appointment time is needed. Assuming uncertainty in decision-making, as in robust Predictive-Reactive Scheduling, strengthens the original schedule and minimises the need for rescheduling.

Production will have already begun before interruptions, according to the principles of dynamic scheduling. Schedule adjustments that do not consider this may need costly changes that result in unnecessary losses. One technique to minimise such unjustified modifications is incorporating the number of jobs that need to adjust the beginning dates originally planned into the target function. Tools created to solve Inverse Optimisation issues are also used for Inverse and Reverse Scheduling issues.

Parameters of a model may be inferred using Inverse Optimisation, whereas unknown parameters for optimum scheduling can be determined using reverse scheduling. Parts of the MSS include a Database Management section, an Automatic Schedule Generator, an Editor for creating and modifying schedules, and an Evaluator for gauging the system's effectiveness. The Automatic Schedule Generator is fed data from the Database Management module. Job scheduling, problem-solving, and looking forward to potential scheduling issues are the primary responsibilities of system users. Production management and efficiency depend heavily on factory scheduling systems. Manufacturers may create scheduling systems that successfully manage and react to the ever-changing production environment if they have a firm grasp of the available methods and the context in which they are most helpful. Traditional hierarchical and centralised systems are replaced by decentralised designs in which autonomous agents interact. Technologies like Cloud Computing, IoT, Big Data, and RFID links facilitate communication between these agents and central decision-making hubs. As IoT includes the digital connection of items to the Internet, cloud computing enables dynamic virtualisation and scaling of resources. In contrast, IoT enables the intelligent integration of devices online and the remote execution of work orders. The term "big data" describes the methods used to handle vast and heterogeneous internet datasets, allowing for finding patterns where none existed before.

Integration of computational and communicative activities with physical processes is made possible by CPS and CPPS, thereby promoting the convergence of real and virtual environments. Like sensor networks, they consist of a web of physical I/O units that may be interacted with. Introducing these new intelligent CPS is expected to increase innovation and competitiveness across various sectors, including aviation, automotive, chemical, energy, infrastructure, and transportation. The CPPS is the offspring of the CPS with manufacturing-specific implementations, and it is made up of autonomous and cooperative parts and subsystems integrated to cover all production process phases, from the shop floor to the logistical networks. The requirement for effective scheduling methods to handle the various and unexpected strains on remote production processes is one of the biggest obstacles faced by these systems. The ANSI/ISA 95 standard is a commonly used model for describing the structure of a company and the hierarchy of authority within it. The Purdue Enterprise Reference Architecture provides a valuable framework for this hierarchical organisation of decision-making, distinguishing between Level 0 (the physical procedure of the production process), Level 2 (the supervision and control of underlying activities), Level 3 (the management of operations and the production workflow to produce the desired products), and Level 4 (the business activities).

The decisions made in CPPS will affect those made in the planning and management of the industry. This is because the CPPS can handle many tasks, from the most fundamental aspects of manufacturing (Level 0) to the strategic oversight of the entire production cycle (Level 3). The resulting dispersed structure, managed jointly by enterprise resource planning (ERP) and CPPS, alters the conventional top-down, centralised approach to manufacturing strategy. Decision-making is shifting to be more decentralised and collaborative, which will have repercussions for scheduling.

Non-permutation programming of fabrication cells is only one example of how incorporating more complicated and flexible production systems may make a significant difference. The scheduling community, facing new issues, must adapt by creating new tactics and approaches. Smart Scheduling is a revolutionary idea that solves the issues brought on by the components' real-time independence in production planning systems. The design of Smart Scheduling is completed by introducing a method to deal with the issues caused by the real-time autonomy of system components. Due to event-driven and trigger-based rescheduling procedures, the Tolerance Scheduling problem is a serious challenge in ERP systems. Predefined criteria relating to the main elements of the event, whether they originate in production resources or tasks to be carried out, are needed to automate this procedure. Even if an incident is determined to fall into the category that prompts rescheduling, its impact may be too little to warrant initiating the procedure.

This introduces a new variable into the mix, increasing the number of potential sources of disruption to the manufacturing process. The method is flexible enough to accommodate various situations regarded as rescheduling triggers. An ideal or near-optimal starting point for this Tolerance Scheduling issue produces a range of tolerances, mainly for the model's parameters. The purpose is to find the schedule value that maximises the distance between the parameters while improving the original objective function up to a specific inertia factor. This gives you a device to see potential rescheduling events and decide whether to proceed.

Using the resources available in Smart Manufacturing and Industry 4.0 ecosystems, we offer Smart Scheduling as a framework for scheduling in production planning. A CPPS is primarily responsible for this. Smart Scheduling aims to provide an automated answer to scheduling issues inside the unified framework of CPPS. In an autonomous CPPS, this method captures the features of a traditional MSS (Machine Structural System) that serves as a member of the MSS-scheduler ensemble. Because of its adaptability and the ability to link to other business processes, the MSS's features lend themselves well to integration into a CPPS. Computer (and physical) systems built using CPPS are connected to facilitate these uses. CPPS is superior to MSS and may be used instead because of its increased capacity for quality evaluations and failure diagnosis.

The Smart Scheduling framework is presented to improve CPPS in this regard, and it is primarily concerned with solving the Tolerance Scheduling problem. Our solution offers robust optimisation-based methods for evaluating the criticality of events concerning their specific characteristics and overall impact. This capability allows the CPPS to improve resilience and reduce susceptibility to environmental noise by delaying rescheduling initiation until the schedule's objectives have been significantly compromised. Smart Scheduling is based on the same principles as dynamic scheduling, in which an initial schedule is generated after solving the issue in its conventional or stochastic form. In the second step, manufacturing begins as planned until an unexpected hiccup is noticed. The occurrence is evaluated to see whether a rescheduling process needs to be initiated. The manufacturing process will proceed if the current schedule is within the tolerance. If this is not the case, a new schedule will be created to account for the change. After a schedule has been generated using the dynamic scheduling approach, fresh tolerances may be established for the Tolerance Scheduling problem.

Finally, Smart Scheduling is a cutting-edge idea that solves problems caused by real-time autonomy in manufacturing planning systems. Smart Scheduling improves production planning and minimises the need for human intervention by using an M scheduler's features. Like an MSS, Smart Scheduling (SS) allows the scheduler to redesign the whole schedule or just a section, depending on the deviation. With this method, businesses may learn from one another's past mistakes and pick effective rescheduling tactics. Tolerance acquired from the solution of the Tolerance Scheduling problem is used by the Smart Scheduling technique to decide whether or not to initiate rescheduling. Based on the dynamic scheduling paradigm, Smart Scheduling maximises the potential of Industry 4.0 and Smart Manufacturing settings. It lets businesses use the knowledge gained across sectors, improving the quality of rescheduling decisions. It also facilitates natural and intuitive decision-making, which is essential for meeting the needs of manufacturing companies and overcoming reluctance to adopt Smart Manufacturing and Industry 4.0 practices.

An automated scheduling process like Smart Scheduling, however, has obstacles. Scheduling relies heavily on data and information management since it uses much information about the production environment. Acceptable solutions need efficient data collection from autonomous and decentralised system parts. Access to the finest available computational techniques is crucial in autonomous and decentralised production settings. By executing routines in high-performance computing facilities, Cloud Computing techniques may expedite calculating solutions to the scheduling issue. Access to these computing resources must be guaranteed to ensure the system's agility and adaptability. Smart Manufacturing and the Industrial Internet of Things (IIoT) place a premium on modularity and scalability. A more difficult-to-manage system results from plug-and-play physical and decisional components. When new parts are integrated into the system, relevant data is made accessible in a manner suitable for decision-making processes.

In conclusion, the Smart Scheduling scheme presents a novel decision-making schema that benefits from the properties of these new contexts and produces adaptable and efficient production schedules on the go. The suggested Smart Scheduling framework employs a powerful screening method (Tolerance Scheduling) to minimise the requirement for rescheduling in the face of unplanned and disruptive occurrences.

2.5.2 Case Study of Experimental Digital Twin

In another work on Streamlining Simulation-based Systems Engineering for Industry 4.0, Schluse et al. (2018) presented experimental digital twins (EDT) in which they explained that a "Technical Asset" (e.g., device, component, environmental feature) might have an identical digital counterpart called a "Digital Twin". Data models, which include geometry and structure, functional models including data processing and behaviour, and communication interface models, are all parts of a Digital Twin. It combines the insights gained from engineering modelling efforts (Digital Model) with operational data collected in the field (Digital Shadow). As a result of the employment of simulators to make the Digital Twin experimental, the concept of an EDT was initially suggested by Schluse et al. (2018).

The authors employed a simulation to depict how EDTs incorporate the concepts of Digital Twins alongside Model-based Systems Engineering as well as simulation technologies to create a novel structuring component for simulation-based systems engineering procedures used in a wide range of contexts, such as development, optimisation, verification, training, actualisation, and the use of user interfaces in intelligent systems. Novel developments in 3D simulation technology make it possible to merge different simulation domains, and simulation technology provides all the essential methodologies for simulating practically any part of complex systems. To systematically investigate the needs, structure, and behaviour of complex systems, Model-based Systems Engineering is being employed in the age of Industry 4.0. The resultant EDT networks integrate and link various model components independently of the specific simulation methodologies. They used their built-in communication features to provide port-based interactions and specialised integration principles to facilitate interaction amongst simulation algorithms. EDTs pave the path for creating intelligent systems using an integrated simulation-based approach, including simulation-based systems engineering, simulation-based optimisation, and simulation-based control.

2.5.3 Case Study of Waste Management

In a similar work by Hasan et al. (2021), which centralised on the application of Industry 4.0 in waste management, tagged "smart waste management", a five-layer framework model was designed to integrate IoT with waste management. These layers include the business layer, application layer, middleware layer, network layer, and perception layer.

2.5.3.1 Network Layer

Expert services are in the perception layer, and the network layer sends the cumulative total to those tiers. The network's communication is a jigsaw puzzle assembled from 6LoWPAN, Bluetooth, GPRS, UMTS, Wi-Fi, Zigbee, and Z-wire components. In addition to its previously described functions, the network layer carries out the cloud services process and the information management operation.

2.5.3.2 Middleware Layer

This middleware layer is a virtualisation technique or many sub-layers that translate communications between different system sections. It is crucial in developing new ideas since it facilitates competition, allowing the application layer to interact with either the perception or relationship layers.

2.5.3.3 Application Layer

While it is true that the application layer does not directly aid in the development of an IoT framework, it is here that many services have crafted an interpretive and accessible graphical user interface.

The working of this smart waste management depends on five dimensions: government dimension, social dimension, economic dimension, technological dimension, and environmental dimension.

The waste disposal model becomes interconnected by developing the proper IoT infrastructure, which uses specific automated technologies to boost the trash service's velocity and profitability. The benefits of the service may, therefore, be maximised by associating the IoT strategy with relevant social, technological, and political variables.

2.6 BLENDING INDUSTRY 4.0 IN THE WATER SYSTEM OPERATIONS (WATER 4.0)

Based on the veracity that water is indispensable to life and that its use in everyday life cannot be over-emphasised, water management becomes imperative. Water is an essential natural resource that helps sustain life, industry, agriculture, the food industry, and the expansion of the nation's economy. Water supply has become more important in recent years, particularly in major cities, due to urban sprawl (the rapid growth of metropolitan areas). Global water use has grown by seven in the previous century, and scientists predict it will increase by another factor of two in the next two decades (Wada & Bierkens, 2014). The water sector is facing a wide range of issues on a global scale (Alabi et al., 2019). One of the significant environmental challenges confronting modern society is the widespread pollution of water supplies by millions of industrial and natural chemical compounds. Over 35% of all fatalities in developing nations are associated with inadequate drinking water, affecting over 3 billion people. According to the WHO, 1.8 million people die yearly from water-related diseases; 88% of these fatalities occur in children living in underdeveloped nations.

Different models project that water consumption worldwide will quadruple during the next two decades, posing a serious risk to public health if adequate measures are not taken to ensure water quality. Both deliberate and accidental pollution of water distribution networks is a serious problem. As a result of these problems, the field of water management has grown in importance and sophistication. There has been progress in water filtration technology, but there is still a long way to go (Sahu et al., 2024). Many utilities are working towards a more digitised and integrated approach to water management to combat these problems. Sensor-based systems and remote monitoring are now being developed to guarantee real-time processing of massive volumes of data and deliver actions for a broad range of management needs. The first generation of systems to monitor the variables necessary for comprehensive qualitative and quantitative water safety relied on human data collected at various network nodes. To satisfy the standards, the European Commission recently introduced the "Water 4.0" concept; these water management systems need increased technical sustainability and secure communications.

The prospect for revolutionary change in water resource management and optimisation is offered by Water 4.0, applying Industry 4.0 concepts and technology to water system operations. Water 4.0 provides various benefits regarding efficiency, sustainability, and resilience in the water industry by integrating cutting-edge digital technology, real-time data analytics, and linked systems. Smart Cities, Smart Enterprises, Smart Factories, Smart Logistics, Smart Manufacturing, and Smart Water are just some of the "Smart Thinking" efforts that have emerged since the introduction of Industry 4.0 in the year 2013 (Alabi et al., 2019). Water 4.0, also known as Smart Water, represents the latest iteration of the Industry 4.0 revolution that the water sector has just adopted. Germany's Water Collaboration conceived the Water 4.0 programme to update the country's antiquated water management and distribution systems for the modern era. This will pave the way for decentralised water management systems. Thanks to the wealth of data, the water sector will make better choices daily, allowing for more consistent demand and supply.

The water supply network covers a vast region, and its many components include water metering, water flexion, disinfection with chlorine, retention, gravity dissemination, and pump delivery. Fundamental needs like energy, water, and gas must be modified to suit the new requirements as the IIoT develops. There will be a need for a uniformly high degree of security across a wide range of communication protocols, standards, and technologies for interconnecting industrial sensors, instruments, control devices, and gearbox units.

2.7 CHAPTER SUMMARY AND CONCLUSION

Industry 4.0 is a significant technological advancement that focuses on creating smart factories that are efficient, flexible, and environmentally friendly. It aims to simplify production processes, maximise resource utilisation, improve product quality, cut costs, and make businesses more competitive. Industry 4.0 encourages interoperability between parts and systems, enabling better communication and collaboration throughout the value chain. Predictive maintenance, data analytics, and real-time data analysis improve productivity and quality. However, Industry 4.0 faces challenges in reinforcing personnel in digital technology, modern management skills, and organisational culture.

Data security and confidentiality are crucial in this age of constant information flow. Collaboration between industry players, governments, academic institutions, and regulatory agencies is crucial for successfully implementing and spreading Industry 4.0 technologies. Key attributes of Industry 4.0 include CPS, IoT, and AI. AI can potentially increase productivity in the manufacturing sector by combining tools that mimic human intelligence and learning. Cloud computing is essential for Industry 4.0 firms, providing near-real-time access to data and proactive management. IoT transforms logistics and shipping by connecting machines directly, enabling location tracking and a cloud-based fleet management solution.

Water management is crucial for everyday life, supporting industries, agriculture, and the economy. Global water use has grown by a factor of seven in the past century and is predicted to increase by another factor of two in the next two decades. Environmental challenges include pollution from industrial and natural chemicals, which affects over 35% of all fatalities in developing nations. Water consumption worldwide is expected to quadruple in the next two decades, posing a serious risk to public health. Utilities are working towards a more digitised and integrated approach, using sensor-based systems and remote monitoring to combat these issues.

Water 4.0, a concept that combines digital technology, real-time data analytics, and linked systems, is adopted by sectors like Smart Cities, Enterprises, Factories, Logistics, Manufacturing, and Smart Water. The Water 4.0 programme aims to update Germany's antiquated water management and distribution systems for the modern era and pave the way for decentralised systems. The government plays a crucial role in providing water for the healthy living of its citizens.

REFERENCES

Adeniran, A. O. (2016). Impacts of the Fourth Industrial Revolution on transportation in the developing nations. *International Educational Scientific Research Journal*, *2*(11), 56–60.

Agarwal, H., & Agarwal, R. (2017). First Industrial Revolution and Second Industrial Revolution: Technological differences and the differences in banking and financing of the firms. *Saudi Journal of Humanities and Social Sciences*, *2*(11A), 1062–1066.

Alabi, M. O., Telukdarie, A., & van Rensburg, N. J. (2019). Water 4.0: An integrated business model from an industry 4.0 approach. In *Proceedings of the IEEE International Conference on Industrial Engineering and Engineering Management* (pp. 1364–1369). GTAI.

Alberto, I. R. I., Alberto, N. R. I., Ghosh, A. K., Jain, B., Jayakumar, S., Martinez-Martin, N., … & Celi, L. A. (2023). The impact of commercial health datasets on medical research and health-care algorithms. *The Lancet Digital Health*, *5*(5), e288–e294.

Bag, S., Telukdarie, A., Pretorius, J. C., & Gupta, S. (2021). Industry 4.0 and supply chain sustainability: Framework and future research directions. *Benchmarking: An International Journal*, *28*(5), 1410–1450.

Beker, C., & Lipsey, R. G. (2002). Science, institutions, and the industrial revolution. Discussion Paper No. 02–4, Department of Economics, Simon Fraser University, Canada.

Bigliardi, B., Bottani, E., Casella, G., Filippelli, S., Petroni, A., Pini, B., & Gianatti, E. (2023). Industry 4.0 in the agrifood supply chain: A review. *Procedia Computer Science*, *217*, 1755–1764.

Birkel, H., Veile, J., Müller, J., Hartmann, E., & Voigt, K.-I. (2019). Development of a risk framework for Industry 4.0 in the context of sustainability for established manufacturers. *Sustainability, 11*, 384. https://doi.org/10.3390/su11020384

Bote-Curiel, L., Munoz-Romero, S., Gerrero-Curieses, A., & Rojo-Álvarez, J. L. (2019). Deep learning and big data in healthcare: A double review for critical beginners. *Applied Sciences*, *9*(11), 2331.

Ceruti, A., Marzocca, P., Liverani, A., & Bil, C. (2019). Maintenance in aeronautics in an Industry 4.0 context: The role of augmented reality and additive manufacturing. *Journal of Computational Design and Engineering*, *6*(4), 516–526.

Choudhary, A. K., & Rao, S. (2018). History of rail transportation and importance of Indian Railways (IR) transportation. *International Journal of Engineering Development and Research*, *6*(3), 73–77.

Clark, G., O'Rourke, K. H., & Taylor, A. M. (2008). New comparative economic history: Made in America? The new world, the old, and the industrial revolution. *American Economic Review: Papers & Proceedings*, *98*(2), 523–528.

Corallo, A., Latino, M. E., & Menegoli, M. (2018). From industry 4.0 to agriculture 4.0: A framework to manage product data in agri-food supply chain for voluntary traceability. *International Journal of Nutrition and Food Engineering*, *12*(5), 146–150.

Deane, P. M., & Cole, W. A. (1962). *British Economic Growth, 1688–1959*. Cambridge University Press.

Ernst, F., & Frische, P. (2015). Industry 4.0/industrial internet of things-related technologies and requirements for a successful digital transformation: An investigation of manufacturing businesses worldwide. Available at SSRN 2698137.

Ferreira, J. J., Lopes, J. M., Gomes, S., & Rammal, H. G. (2023). Industry 4.0 implementation: Environmental and social sustainability in manufacturing multinational enterprises. *Journal of Cleaner Production*, *404*, 136841.

George, A. S., & George, A. H. (2020). Industrial Revolution 5.0: The transformation of the modern manufacturing process to enable man and machine to work hand in hand. *Journal of Seybold Report*, ISSN No, 1533, 9211.

Groumpos, P. P. (2021). A critical historical and scientific overview of all industrial revolutions. *IFAC-PapersOnLine*, *54*(13), 464–471.

Hasan, M. A., Raghuveer, K., Pandey, P. S., Kumar, A., Bora, A., Jose, D., … & Khanapurkar, M. M. (2021). Internet of things and it's application in Industry 4.0 for smart waste management. *Journal of Environmental Protection and Ecology*, *22*(6), 2368–2378.

Hobsbawm, E. (2016). *Da Revolução Industrial inglesa ao imperialismo*. 6ª. Ed. Forense Universitária.

Javaid, M., Haleem, A., Singh, R. P., Suman, R., & Gonzalez, E. S. (2022). Understanding the adoption of Industry 4.0 technologies in improving environmental sustainability. *Sustainable Operations and Computers*, *3*, 203–217.

Kagermann, H., Wahlster, W., & Helbig, J. (2013). Securing the future of German manufacturing industry. *Recommendations for Implementing the Strategic Initiative INDUSTRIE 4.0*. Final Report of the Industrie 4.0 Working Group, Acatech— National Academy of Science and Engineering, 678 p.

Karnouskos, S., Colombo, A. W., Leitao, P., & Ribeiro, L. (2021). Guest Editorial Industrial Agents: Concepts, Technologies, and Applications. *IEEE Journal of Emerging and Selected Topics in Industrial Electronics,* 3(1), 2–4.

Klingenberg, C., & Antunes, J. (2017). Industry 4.0: What makes it a revolution. *EurOMA, 2017*, 1–11.

Kurt, R. (2019). Industry 4.0 in terms of industrial relations and its impacts on labour life. *Procedia Computer Science*, *158*, 590–601.

Landes, D. S. (1969). *The Unbound Prometheus: Technological Changes and Industrial Development in Western Europe from 1750 to Present*. Cambridge University Press. ISBN 978-0-521-09418-4.

Lee, E. A. (2008, May). Cyber physical systems: Design challenges. In *2008 11th IEEE International Symposium on Object and Component-Oriented Real-Time Distributed Computing (ISORC)* (pp. 363–369). IEEE.

Lehmann-Hasemeyer, S., Prettner, K., & Tscheuschner, P. (2023). The scientific revolution and its implications for long-run economic development. *World Development*, *168*, 106262.

Mavropoulos, A., & Nilsen, A. W. (2020). *Industry 4.0 and Circular Economy: Towards a Wasteless Future or a Wasteful Planet?* John Wiley & Sons.

Mohajan, H. K. (2019). The first industrial revolution: Creation of a new global human era. *Journal of Social Sciences and Humanities, 5*, 377–387.

Myers, B., Hudson, S. E., & Pausch, R. (2000). Past, present, and future of user interface software tools. *ACM Transactions on Computer-Human Interaction (TOCHI)*, *7*(1), 3–28.

Nuvolari, A. (2019). Understanding successive industrial revolutions: A "development block" approach. *Environmental Innovation and Societal Transitions*, *32*, 33–44.

O'Brien, P. (2017). The contributions of warfare with revolutionary and Napoleonic France to the consolidation and progress of the British industrial revolution: Revised version of working paper 150.

Oztemel, E., & Gursev, S. (2020). Literature review of industry 4.0 and related technologies. *Journal of Intelligent Manufacturing, 31*, 127–182.

Pandey, V., Sircar, A., Bist, N., Solanki, K., & Yadav, K. (2023). Accelerating the renewable energy sector through Industry 4.0: Optimization opportunities in the digital revolution. *International Journal of Innovation Studies*, *7*(2), 171–188.

Pearson, R., & Richardson, D. (2001). Business networking in the industrial revolution. *Economic History Review*, *54*(4), 657–679.

Psarommatis, F., May, G., & Azamfirei, V. (2023). Envisioning maintenance 5.0: Insights from a systematic literature review of Industry 4.0 and a proposed framework. *Journal of Manufacturing Systems*, *68*, 376–399.

Rahman, N. S. F. A., Hamid, A. A., Lirn, T. C., Al Kalbani, K., & Sahin, B. (2022). The logistics industry adoption of Industry 4.0 practices: A systematic review of the gulf region. *Cleaner Logistics and Supply Chain*, *5*, 100085.

Rossit, D. A., Tohmé, F., & Frutos, M. (2019). Industry 4.0: Smart scheduling. *International Journal of Production Research*, *57*(12), 3802–3813.

Roy, S., Meena, T., & Lim, S. J. (2022). Demystifying supervised learning in healthcare 4.0: A new reality of transforming diagnostic medicine. *Diagnostics*, *12*(10), 2549.

Sahu, S., Supe, J. and Rathore, K., 2024. Assessment of an Advanced Water Filtration System for Enhanced Water Purification. *International Journal for Multidisciplinary Research*, *6*(5).

Schluse, M., Priggemeyer, M., Atorf, L., & Rossmann, J. (2018). Experimentable digital twins—Streamlining simulation-based systems engineering for industry 4.0. *IEEE Transactions on Industrial Informatics*, *14*(4), 1722–1731.

Sharma, A., & Singh, B. J. (2020). Evolution of industrial revolutions: A review. *International Journal of Innovative Technology and Exploring Engineering*, *9*(11), 66–73.

Sugden, K., & Cockerill, A. (2017). The wool and cotton textile industries in England and Wales up to 1850. In L. Shaw-Taylor, A. Cockerill, & M. Satchell (Eds.), *The Online Historical Atlas of Occupational Structure and Population Geography in England and Wales* (pp. 1–16, 1600–2011). Cambridge University Press.

Viollet, P. L. (2017). From the water wheel to turbines and hydroelectricity. Technological evolution and revolutions. *Comptes Rendus Mécanique*, *345*(8), 570–580.

Wada, Y., & Bierkens, M. F. (2014). Sustainability of global water use: past reconstruction and future projections. *Environmental Research Letters*, 104003.

Wang, L., Von Laszewski, G., Younge, A., He, X., Kunze, M., Tao, J., & Fu, C. (2010). Cloud computing: A perspective study. *New Generation Computing*, *28*, 137–146.

Weightman, G. (2010). *The industrial revolutionaries: The making of the modern world 1776–1914.* Grove Press. ISBN 978-0-8021-4484-3.

Weightman, G., 2018. *The Industrial Revolutionaries: The making of the modern world, 1776–1914.* Open Road+ Grove/Atlantic.

Zhang, C. and Lu, Y., 2021. Study on artificial intelligence: The state of the art and future prospects. *Journal of Industrial Information Integration*, *23*, 100224.

3 Water 4.0
Perspective and Concepts

3.1 INTRODUCTION

Water is essential to humans, plants, and animals. It is also an essential part of every nation's economy since it is part of many production sectors of the world. Everyone has a part in the water flow of this world. Hence, the government must make provisions for water processing and distribution in the country for the healthy living of its citizens. Nevertheless, water is useless without men because there would be no one to use it, so man without water cannot live longer, as said earlier. History made us understand that the processing and distribution of water started in ancient times, originating in Greece (c. 32000–1100 B.C.) (Mays et al., 2012). However, popularly, the credit was mainly given to Rome because it built the first water technology, laying a foundation for other water technologies (Water 1.0–4.0). History spanned through to the 16th century, when the invention of the steam engine began to power the force of pumping out water from rivers, oceans, etc.

With Water 1.0, the honour of the inventors was given to the Roman Empire, which made the innovation very simple. It was said that they dealt with the importation of water, and as such, they created an innovation of a pipe that runs through from the river into individual homes for use. The end came to this innovative thinking with the fall of the Roman Empire, as it is already known that nothing survives once a fall comes upon a nation. With this fall, other kingdoms that benefited from this innovation took another means of producing water for themselves. The next in line was Water 2.0, which came with drinkable water, although with some next-level innovations and some reasons backing it up. One of the reasons Water 2.0 came into existence was the cholera outbreak in the United States, and this gave birth to the action of some scientists and engineers to come up with water purification processes for drinking. This stirred up the use of steam engines to pump the water for consumption. Water 3.0 never came on good terms, according to Sedlak (2014). It was just based on the treatment of contaminated water, in which the engineers had to work tremendously to export contaminated water back to rivers for the people to live healthily. The engineers and the scientists had used water treatment "Microbes", and it is said that the microbes also had their effects, but that is for another day. Regarding Water 4.0, which we are dealing with in this book, it is assumed that the approach will be in two ways. The first approach is that Water 4.0 could take up the leftover of Water 3.0 and continue with that. Secondly, it could begin with the implementation of Water 1.0, which came from the Roman Empire but is more advanced and a process in which homes will be self-reliant of their own accord (see Chapter 2 for further details).

DOI: 10.1201/9781003633839-3

Water 4.0, in this respect, is known to be the future of water, “Smart Water” or “Digital Water”, which we are already in. In this chapter, we will look into what Water 4.0 looks like and what it entails. We would note that there is what is called Industry 4.0, which is the fourth revolution in the manufacturing processes. The application of water is what is known as “Water 4.0”. As said in the initial paragraph, stating that water is life, there is also a need for us to note that there is no need for humans to labour before getting water to drink or do anything else. As technology is ravaging all other sectors of living, there is a need for innovative technologies to be applied in the water sector, since it is “life”. Starting from the purification of the water for drinking and other usages of water, there is a need for Industry 4.0 to come into place, which is the combination of artificial intelligence (AI) and machines known as the “Cyber-Physical System” (CPS) (its benefits have been stated in Chapter 2). As such, Water 4.0 is not just a standard word or phrase coined out of something but a concept.

Stating the usefulness of CPSs, which are efficient and productive, fast and safe, best in decision-making, etc., they can be applied to water supply and management also. Owing to the veracity that water is being used in almost every sector of life, especially in agriculture to grow plants, water is needed. With the use of CPSs, it makes it possible for a farmer to water his farm without even getting to that farm. When there is automated irrigation for watering the farm, the farmer waters his farmland for cultivation and productivity without stress. In decision-making processes, the CPS had to be programmed in water systems without manual operations while delivering, distributing, and supplying water to its destination.

3.2 WATER 4.0: PERSPECTIVE AND CONCEPTS

Water 4.0 is an approach to water management that emphasises digitisation and automation to be more efficient, adaptable, and competitive with scarce water resources. Water 4.0 emphasises the integration of equipment, techniques, storage devices, and operation resources, as well as the Internet of Things (IoT), smart grids, and services. Water 4.0 serves as a catch-all phrase that organises these components into a hierarchical structure relevant to water management. CPS provide optimum connectivity between virtual and actual water systems in Water 4.0, where software tools are employed in designing, building, and managing these systems. This will make room for a comprehensive strategy across every phase of the value chain by facilitating the development of an intelligent network that connects water users (i.e., agriculture, industry, and households) and components within a sustainable water infrastructure. Water 4.0 also offers high transparency for water consumers, meeting present demands while presenting future possibilities (Kijak, 2022).

There are different applications of Water 4.0, including pipeline health monitoring, pressure control and monitoring, water quality monitoring, and leakage detection and monitoring. Since the distribution systems for water are critical for delivering sanitary water to end consumers, monitoring pipeline health is crucial. Damaged pipes or those broken by outside forces may cause severe water loss. To minimise losses and avoid deformation, tracking pipe health beforehand is essential. Technologies, including accelerometers, wireless sensor networks, and soft capacitive sensors, are utilised to keep tabs on pipelines. Utilities may enhance system operations using

Water 4.0 technology, which can immediately notify deformation regions and active monitoring mechanisms. An example of a robotics-based pipe surveillance system is the use of smart robots. Using AI at the network's periphery might lead to fully autonomous, closed-loop systems for monitoring vital infrastructure and assuring the reliable functioning of water distribution networks (WDNs).

Pressure sensors detect water pressures in pipes and at network nodes in water distribution systems. A positive correlation exists between pressure and freshwater demand. More pressure is needed to keep up as demand rises. However, leaks may make pressure management more difficult, since customer demand is often not linear. Efficient pressure control in networks is also impacted by demand unpredictability. With the help of new technology, pressure in leaky pipes may be actively controlled by monitoring pressure and demand at nodes in real-time. Monitoring parameters, including dissolved oxygen content, residual chlorine in WDNs, integrated sensor devices, temperature, pH, and turbidity, are necessary for water quality applications. The ideal choice for determining water status in distribution pipe networks is a combination of Water 4.0 characteristics, active decision support systems, and control valves. The pollution of water systems can only be stopped in its tracks by swift and decisive intervention. It may be possible to close valves at the node to which the tainted water is linked or at the district-metered area's entry point to stop contamination from spreading. For pipe leakage monitoring, a system with integrated sensors for measurement and control includes acoustic, flow sensors and actuators, pressure, and vibration. Water 4.0 uses machine learning to find leaks in the system and offers constant, real-time monitoring to stop further water loss. This technology allows for proactive leak detection, correction, and prevention.

The IoT, AI, and other innovative technologies fuelled a 5.11% compound annual growth rate (CAGR) in the water utility monitoring market between 2018 and 2022. These innovations can potentially revolutionise the water industry by enhancing operational water management and fixing the underlying causes of water waste. At a CAGR of 13.5% from 2019 to 2024, the utility IoT market is projected to increase from an initial 2019 estimate of $28.6 billion to $53.8 billion. Real-time monitoring of complicated water supply distribution networks is complex in any case. Water 4.0 is a systematised approach to maximising the efficiency of WDNs and enhancing their capacity for real-time monitoring.

Safeguarding water supply networks, preventing unauthorised access, and supplying security monitoring systems are all examples of how information and communication technology (ICT) may help address these issues. On the other hand, wireless system security vulnerabilities may pose threats to the water supply system's (WSS) infrastructure surveillance. Denial-of-service, distributed denial-of-service, and advanced persistent denial-of-service assaults are common in ICT-based water management systems, resulting in lengthy service interruptions and financial losses for utilities. The proliferation of IoT has resulted in collecting massive volumes of information specific to the use context. AI approaches can only function as intended if the inherent uncertainty in Big Data (BD) analytics is reduced. Constant power supply for CPS or IoT sensing and communication devices is required for continuous and autonomous monitoring of WSS infrastructure, and improved battery life and better energy utilisation are necessary for applications.

3.3 KEY FEATURES OF WATER 4.0

Industry 4.0, often known as the Fourth Industrial Revolution, has been fuelled by contemporary technologies, including AI, IoT, and big data (Wan et al., 2021). Industry 4.0, which captures the present shift in business and industry towards more intelligence, demonstrates how these technologies are quickly changing the industrial landscape. This paradigm change is caused by quick technological advancements and the rising demand for these technologies in industrial applications (Lasi et al., 2014; Foivos et al., 2023).

The concept has been used in other engineering-related disciplines, including medicine, smart logistics, and the energy industry (Minoli et al., 2017; Bedi et al., 2018). Water 4.0 is the term used to describe the application of these technologies in the water industry. The earliest revolution in the water industry was piping potable water into urban areas and sewage out of them using improvised local systems. During the second stage, a sizable, centralised infrastructure was used to treat drinking water to eliminate infectious germs and safeguard people from diseases spread by the very success of Water 1.0. Computers and control were widely used in the water industry when it reached version 3.0. Water 4.0, the current stage of the water revolution, calls for an integrated approach to using cutting-edge digital technologies, such as cloud computing, AI, and automation, to improve WSS operations (Adedeji et al., 2022). Barroso et al. (2023) use the IoT to create a smart campus, as described in their article under the moniker SmartPoliTech. The primary emphasis of their initiative is to develop a sustainable water management system capable of predicting water usage and reducing water waste.

3.3.1 Internet of Things

The industrial sector has experienced enormous growth, thanks to the IoT, which was seen as a critical technology for creating the next wave of manufacturing systems. The IoT allows industrial managers to automatically collect and analyse data to improve decisions and maximise productivity. Factory-level IoT connectivity solutions transmit data from sensors and machines to the cloud. Systems for providing water can use the same technology. Supervisory control and data acquisition (SCADA) systems are now used to manage the functioning of WSSs.

The IoT is expected to offer good revolutionary options in Water 4.0 to enhance the performance of numerous industrial technologies, including SCADA. Sensing devices connected to the Internet and smart water networks provide relevant information on the condition of WDNs to aid in managing, treating, and decision-making. Because of IoT smart sensors and gadgets, complex water piping networks may be continuously monitored in real-time. With IoT, leakage, circulation, water flow, water level, and pressure along the distribution system can be successfully monitored in real-time. The production industry has experienced significant growth, thanks to the IoT, which was viewed as a critical technology for transforming the present manufacturing systems for the next generation of manufacturing systems (Varghese & Tandur, 2014; Trappey et al., 2017; Alabi et al., 2019).

Smart factories can now autonomously collect and analyse data because of the IoT, allowing them to make more informed choices and maximise productivity.

Factory IoT connection solutions send collected information from sensors and equipment to remote servers. WDNs may also use a similar technique (Adedeji et al., 2022). Additionally, the intricate distribution system may provide real-time water quality monitoring. This is a step in the direction of managing water networks more intelligently. Many nations are investing in "smart water network" projects. To strengthen smart water networks and cut utilities' water loss to less than 12%, governments like Singapore, South Korea, and Malta have instituted regulatory laws requiring digital technology adoption. Another recent example of digital technology in the water industry is smart metering, which uses automated meter reading and advanced meter infrastructure (AMI). Accurate metering and billing are achieved via deploying AMI networks. These gadgets can lessen the financial burden brought on by water leaks and increase the reliability of usage-based pricing (Trappey et al., 2017; Adedeji et al., 2022).

Many cities and towns are already spending money on AMI for their water services. A MICROmega Group company in South Africa has introduced the utiliMeter. This AMI-enabled water management device can be connected to a conventional water meter to create an intelligent, prepaid water metering system that meets standard transfer specification requirements. This technique is possible for prepaid, post-paid, flat rate, and flow-limiting water metering (Adedeji et al., 2022).

At the close of 2017, 24,000 homes in South Korea had smart water meters installed by Gochang Water Works. These can control the water supply via active data transfer and real-time communication. The use of complete package sealing technology achieves IP68 protection. Using Vodafone's Narrowband-IoT, Southeast Water and industry specialists have launched experiments to create and link smart water meters and to deploy acoustic sensors along subterranean pipes. Therefore, IoT is growing in popularity in the water industry (Trappey et al., 2017; Adedeji et al., 2022).

3.3.2 Cyber-Physical System

The CPS is the foundation of Industry 4.0, a fusion of networks, computation, and the physical environment. CPSs enable embedded computing devices to sense, monitor, and control the physical environment continuously, transforming the development of ICT (Varghese & Tandur, 2014; Kim, 2017). In Water 4.0, the physical process of a CPS can include the entire WSS or a section of the system, such as monitoring water quality, leakage flows, or energy consumption (Haque et al., 2014; Monostori, 2014). Different sensors and actuator mechanisms, such as pressure, flow, water quality, accelerator sensors, and control valves, are deployed for real-time measurement and control of the physical environment (Taormina et al., 2017).

CPSs are similar to emerging research topics like IoT, machine-to-machine systems, and ubiquitous, fog, and pervasive computing but with better control capabilities. Standard water quality and anomaly detection sensors include pH, dissolved oxygen concentration, flow rate, and turbidity sensors. Pressure, flow rate, acoustic, ultrasonic, and temperature sensors are frequently used for leak detection. A combination of these sensors is employed for leak detection. Temperature sensors provide continuous measurements of the outside temperature within the pipe environment, which are used to create a baseline. A deviation from the baseline indicates the presence of a leak. The CPS is closely related to other emerging research topics like IoT, machine-to-machine systems, and ubiquitous computing, but with better control capabilities (Adedeji et al., 2022).

Ahmed et al. (2017), in their study on water distribution tagged "WADI: A water distribution testbed for research in the design of secure cyber-physical systems" which aimed at conducting research on the design of safe CPSs in WDNs, described that the testbed is put to use for doing security analyses, conducting experiments on detection methods for cyber and physical assaults, and researching the cascade effects that attacks have on CPSs that are linked to one another. It offers a framework for conducting security analysis, testing systems for attack detection, and researching the effects of assaults on linked CPSs.

They utilised a water distribution testbed named WADI. The testbed is used to research security, try different detection techniques for cyber and physical assaults, and investigate the impact of attacks cascading across linked CPSs. WADI's architecture and communication infrastructure are broken out here, along with its many pieces of hardware and software. Experiments that are carried out on WADI include initiating assaults, such as faking level sensor readings, to interrupt the flow of water to consumer tanks. The need to develop cutting-edge testbeds for WDNs is another topic covered in their research. Additionally, the publication emphasises the significance of testbeds for advancing research on CPS security. The study's finding is that WADI, a water distribution testbed, may be used to explore the design of safe CPSs in WDNs.

A study by Alexandra et al. (2023) reviews CPSs' usage in water management and governance and discusses their ramifications. The CPS, also known as "smart" technologies, uses AI and automation to improve the monitoring and administration of water systems. The need to incorporate social and ecological considerations into CPS research and design is emphasised throughout this book. The goal of the document is to achieve sustainable water management. It investigates several CPS tasks, including sensing, data management, decision-making, community, and machine learning. This analysis provides a landscape-scale viewpoint, looking at the uses of CPSs in various water-related situations and industries. The authors advocate for a cybernetic approach that considers social and ecological objectives within water governance. They highlight the need for a comprehensive and system-wide assessment of CPSs in the water field to understand their benefits and challenges. In general, this book highlights the revolutionary role that CPSs may play in water management and the relevance of sustainability in their creation and implementation.

3.3.3 Intelligent Computing

Soldevila et al. (2022) integrate expert knowledge with data-driven technologies to detect and locate water leaks in WDNs. The effectiveness of the proposed strategy to improve leak detection and localisation was tested using a data set of actual and simulated leaks from the Barcelona Water Distribution Networks. In Xiang et al. (2021), we see the development of a smart, dynamic, and flexible strategy for water resource planning. Quinones-Grueiro et al. (2021) proposed using a computationally intelligent system to identify and localise leaks in WDNs. According to the data, this hypothetically intelligent system adequately identifies leaks, estimates their sizes, and pinpoints their locations (Adedeji et al., 2022). The evidence for this comes from the WDN in Modena.

Research by Kowalski et al. (2022) uses intelligent computing by implementing a quasi-intelligent diagnostic approach for a WSS. A numerical model and a SCADA monitoring system are included in the technique, and the system is set up to make consistent comparisons between the monitoring data and the simulation findings. Technology can identify aberrant functioning of the WSS, such as leakages and water outages, and the obligation to establish compliance limitations. When certain thresholds are breached, warnings are created to notify the operators. Incorporating intelligent computing into the system enables it to reduce operational costs while ensuring that clients get the appropriate amount of water. In addition, for effective monitoring, the technique uses a Geographic Information System (GIS) database, a remote water meter-reading device, and an expanded number of sensors. This properly determines the amount of water lost and locates any leaks, giving the dispatchers a specialised informatics tool that will allow them to manage the WSS properly. In general, this technology illustrates how intelligent computing may be used to improve the efficiency of the operation of the WSS.

3.3.4 Cloud Computing

To provide consumers with a secure and high-quality service, cloud computing is one of the emerging technologies (Wang et al., 2010). Technologies based on the cloud have the potential to greatly aid in the implementation of Water 4.0. The cloud's data storage, computation, and service capacity are formidable (Patil et al., 2012). With the IoT data-gathering potential, utilities may use scalable, on-demand services made possible by the Internet.

Water 4.0 plans to use cloud computing to pool resources like water pipe network data to achieve consistency and economies of scale. WDNs, water treatment facilities, and pumping stations contribute to the massive amounts of big data produced by WSSs. Monitoring WSSs is simplified because a large amount of data may be transferred to a cloud computing centre for processing, calculation, and storage. Many businesses now depend on cloud computing, storage, and networking architecture because of its convenience and scalability for AI training and conclusions. Water quality monitoring using the CPS has seen widespread application, with sensor technologies collecting vast amounts of data to track variations in water quality indicators throughout a pipeline system. This is often accomplished via sensors, wireless networks, and actuator networks for control. Although the CPS has superior control capabilities, the IoT has been extensively employed for similar purposes (leak detection and monitoring, water quality detection, pipe health monitoring, etc.) (Taormina et al., 2017; Adedeji et al., 2022).

The use of AI in the water sector is now the topic of several academic investigations. To do this, you must collect relevant information about the utility facility, analyse that information, and use a decision support system to arrive at the best possible conclusions. In this case, we use machine learning techniques from both the supervised and unsupervised categories. Some of the most well-known uses of AI in the water industry include identifying anomalies and predicting future water needs. However, it cannot be disregarded that it is used in water treatment plants for pipe health monitoring and water quality detection. Cloud computing is one of these digital technologies. However, it is seldom employed in the water industry. The management of water resources and the detection of water quality are only two areas where this all-encompassing method has been used (Adedeji et al., 2022).

3.3.5 Automation and Control

Digital technologies for optimising water network operation tasks may be used to create long-term, comprehensive strategies for holistic water network management. This will act as the central control and master database for all water facilities. To handle water supply operations more effectively, one of the new technologies needed for Water 4.0 is AI. Several projects are underway to create algorithms that intelligently plan water resources' dynamic management utilising smart sensors and artificial neural networks. Using AI techniques, we may install robotic sensors in water systems, construct prediction models for future water usage, and integrate blockchain technology-related financial operations related to meter billing (Adedeji et al., 2022).

The potential advantages of digital technology and automation in the water industry, particularly in water delivery, are discussed in Savić (2022). It brings to light that automation and digital transformation could usher in substantial shifts in the protection, control, and operation of complex water systems. Despite this, their work stressed that "humans in the loop" are still necessary for automation in the water industry. Despite the progress made in automation technology in other sectors, such as the automotive and aerospace industries, the safe operation of water systems continues to depend on a workforce that has a high level of expertise.

In addition, human operators who manage water systems need instruction in digital technology systems. To effectively deploy and operate automation systems, it is vital to have a thorough understanding of the systems' complex nature and their inherent constraints. Overconfidence in the capabilities of automation systems may lead to significant failures, as was observed in instances that occurred in the automobile and aviation manufacturing sectors. Overall, the findings of this study reveal that, even though automation and digital transformation can considerably enhance WSSs, human operators must continue to play a vital role in guaranteeing these systems' safe and efficient functioning.

3.4 CHAPTER SUMMARY AND CONCLUSION

Water 4.0 pertains to integrating digital technologies to enhance utilisation, minimise the efficiency of water utilisation, and minimise expenditures associated with water management practices. This innovation, called Smart Water, is widely regarded as the forthcoming paradigm shift in urban water systems.

Water 4.0 has the potential to contribute to environmental conservation through the optimisation of water utilisation and reduction in expenditures associated with water management. This, in turn, can facilitate the adoption of more sustainable practices in utilising water resources. Nevertheless, the environmental advantages of Water 4.0 are not inherently linked to mitigating greenhouse gas emissions or other environmental consequences. Some smart tools are available across developed nations in water management sectors to control, analyse, and monitor day-to-day activities.

REFERENCES

Adedeji, K. B., Ponnle, A. A., Abu-Mahfouz, A. M., & Kurien, A. M. (2022). Towards digitalization of water supply systems for sustainable smart city development—Water 4.0. *Applied Sciences*, *12*(18), 9174.

Ahmed, C. M., Palleti, V. R., & Mathur, A. P. (2017, April). WADI: A water distribution testbed for research in the design of secure cyber physical systems. In P. Tsakalides & B. Beferull-Lozano (Eds.), *Proceedings of the 3rd International Workshop on Cyber-Physical Systems for Smart Water Networks* (pp. 25–28).

Alabi, M. O., Telukdarie, A., & Jansen van Rensburg, N. (2019). *Industry 4.0: Innovative solutions for the water industry*. In E. Schott, H. Keathley, & C. Krejci (Eds.), *Proceedings of the American Society for Engineering Management 2019 International Annual Conference* (ASEM 2019).

Alexandra, C., Daniell, K. A., Guillaume, J., Saraswat, C., & Feldman, H. R. (2023). Cyber-physical systems in water management and governance. *Current Opinion in Environmental Sustainability*, *62*, 101290.

Barroso, S., Bustos, P., & Nunez, P. (2023). Towards a cyber-physical system for sustainable and smart building: A use case for optimising water consumption on a smartcampus. *Journal of Ambient Intelligence and Humanized Computing*, *14*(5), 6379–6399.

Bedi, G., Venayagamoorthy, G. K., Singh, R., Brooks, R. R., & Wang, K. C. (2018). Review of internet of things (IoT) in electric power and energy systems. *IEEE Internet of Things Journal*, *5*, 847–870.

Haque, S. A., Aziz, S. M., & Rahman, M. (2014). Review of cyber-physical system in healthcare. *International Journal of Distributed Sensor Networks*, *10*(4), 217415.

Kijak, R. (2022). Water 4.0: Enhancing climate resilience. In *The Palgrave Handbook of Climate Resilient Societies* (pp. 435–473). Springer International Publishing.

Kim, J. H. (2017). A review of cyber-physical system research relevant to the emerging IT trends: Industry 4.0, IoT, big data, and cloud computing. *Journal of Industrial Integration and Management*, *2*(03), 1750011.

Kowalski, D., Kowalska, B., & Suchorab, P. (2022). Smart water supply system: A quasi intelligent diagnostic method for a distribution network. *Applied Water Science*, *12*(6), 135.

Lasi, H., Fettke, P., Kemper, H. G., Feld, T., & Hoffmann, M. (2014). Industry 4.0. *Business & Information Systems* Engineering, *6*, 239–242.

Mays, L. W., Sklivaniotis, M., & Angelakis, A. N. (2012). Water for human consumption through history. In A. N. Angelakis, L. W. Mays, D. Koutsoyiannis, & N. Mamassis (Eds.), *Evolution of Water Supply through the Millennia*. IWA Publishing. *(7) (PDF) A History of Water Distribution Systems and their Optimisation*. Available from: https://www.researchgate.net/publication/274699540_A_History_of_Water_Distribution_Systems_and_their_Optimisation [accessed Aug 01, 2023].

Minoli, D., Sohraby, K., & Occhiogrosso, B. (2017). IoT considerations, requirements, and architectures for smart buildings—Energy optimization and next-generation building management systems. *IEEE Internet of Things Journal*, *4*, 269–283.

Monostori, L. (2014). Cyber-physical production systems: Roots, expectations and R&D challenges. *Procedia CIRP*, *17*, 9–13.

Patil, V. C., Al-Gaadi, K. A., Biradar, D. P., & Rangaswamy, M. (2012). Internet of Things (IoT) and cloud computing for agriculture: An overview. *Proceedings of Agro-Informatics and Precision Agriculture (AIPA 2012), India*, *292*, 296.

Quinones-Grueiro, M., Milián, M. A., Rivero, M. S., Neto, A. J. S., & Llanes-Santiago, O. (2021). Robust leak localization in water distribution networks using computational intelligence. *Neurocomputing*, *438*, 195–208.

Savić, D. (2022). Digital water developments and lessons learned from automation in the car and aircraft industries. *Engineering*, *9*, 35–41.

Sedlak, D. (2014). *Water 4.0: The Past, Present, and Future of the World? S Most Vital Resource*. Yale University Press.

Soldevila, A., Boracchi, G., Roveri, M., Tornil-Sin, S., & Puig, V. (2022). Leak detection and localization in water distribution networks by combining expert knowledge and data-driven models. *Neural Computing and Applications*, *34*, 1–21.

Taormina, R., Galelli, S., Tippenhauer, N. O., Salomons, E., & Ostfeld, A. (2017). Characterizing cyber-physical attacks on water distribution systems. *Journal of Water Resources Planning and Management*, *143*(5), 04017009.

Trappey, A., Trappey, C., Govindarajan, U., Chuang, A., & Sun, J. (2017). A review of essential standards and patent landscapes for the internet of things: A key enabler for industry 4.0. *Advanced Engineering Informatics*, *33*, 208–229.

Varghese, A., & Tandur, D. (2014). Wireless requirements and challenges in industry 4.0. In *Proceedings of the IEEE International Conference on Contemporary Computing and Informatics* (pp. 634–338), Mysore, India, 27–29.

Wan, J., Li, X., Dai, H. N., Kusiak, A., Martinez-Garcia, M., & Li, D. (2021). Artificial-intelligence driven customized manufacturing factory: Key technologies, applications, and challenges. *Proceedings of the IEEE*, *109*, 377–398. https://doi.org/10.1109/JPROC.2020.3034808

Wang, L., Von Laszewski, G., Younge, A., He, X., Kunze, M., Tao, J., & Fu, C. (2010). Cloud computing: A perspective study. *New Generation Computing*, *28*, 137–146.

Xiang, X., Li, Q., Khan, S., & Khalaf, O. I. (2021). Urban water resource management for sustainable environment planning using artificial intelligence techniques. *Environmental Impact Assessment Review*, *86*, 106515.

4 IoT in the Water Supply Network Management

4.1 INTRODUCTION

The Internet of Things (IoT) is a network of physical objects, such as sensors, meters, and controllers, linked to the Internet and can communicate with one another and other systems. Examples of such devices are sensors, meters, and controllers that are linked to the Internet. IoT technology can gather and communicate real-time data on water quality, flow rate, pressure, and temperature, among other factors, in the context of managing the water supply network. Operators may monitor the system remotely, automate specific tasks, and make educated decisions using data analysis via the integration of IoT technology into the water supply network. This can lower operational expenses, speed up responses, and increase the system's overall dependability and sustainability. For instance, Bandyopadhyay and Sen (2015) proposed an IoT-based system for monitoring water quality that uses sensors to gather data on the physical and chemical characteristics of the water and transmits the information to a cloud-based server for analysis. Akin to this, Kumar et al. (2017) created an IoT-based innovative water management system that employs sensors and actuators to continuously monitor and regulate water supply networks. Water management could change because of the adoption of IoT technology in water supply networks since it will make decisions easier, reduce water waste, and boost efficiency. The use of IoT is accelerating in the transportation, energy, and agriculture industries in Europe, one of the world's largest and strongest IoT marketplaces. The Digital Single Market objective of the European Union (EU) has a big impact on IoT uptake in Europe. The policy aims to remove regulatory barriers and promote regional digital technology development. Additionally, the EU has launched initiatives to support the development of IoT, including the IoT European Large Scale Pilots Programme and the Alliance for Internet of Things Innovation. A few European nations have also started implementing their plans to aid in developing the IoT, including funding for R&D, the creation of testbeds, and innovation hubs (Kagermann et al., 2013). The ambition for digital transformation, automation, and data analytics is expected to fuel an additional rise in IoT adoption in Europe in the coming years. According to Kagermann et al. (2013), other difficulties must be resolved, including requiring qualified people to develop and maintain IoT systems, data privacy and security concerns, and interoperability problems. IoT can also help with early problem detection, enabling rapid corrective action and enhancing the system's reliability and sustainability.

IoT (see Figure 4.1) is becoming more widely used in North America, substantially impacting several businesses and applications.

IoT is widely adopted in the United States across various sectors, including manufacturing, healthcare, transportation, energy, and agribusiness. The National Institute

DOI: 10.1201/9781003633839-4

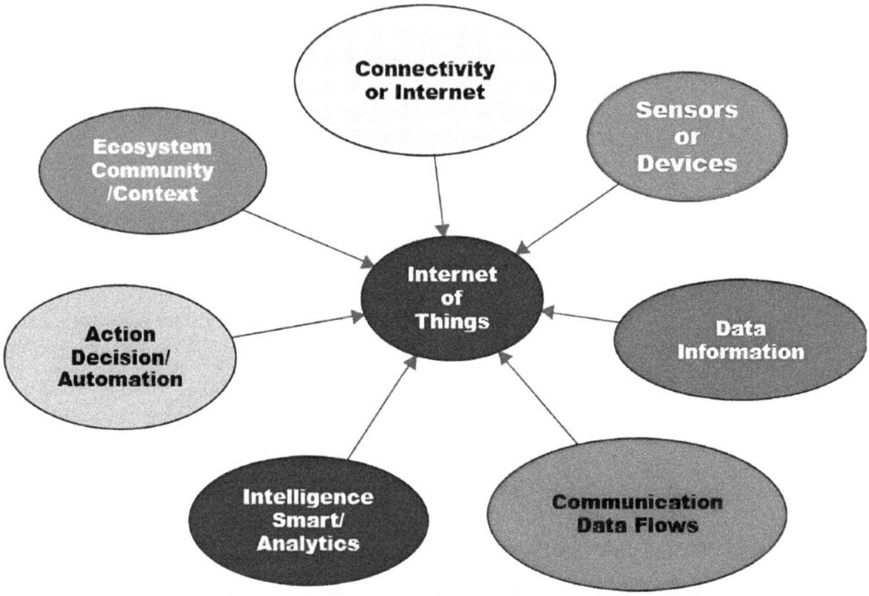

FIGURE 4.1 Conceptual representation of Internet of Things (IoT) integration within Water 4.0 frameworks, emphasising interconnected sensor networks, real-time data acquisition, and advanced analytics for optimised water resource management and infrastructure resilience.

of Standards and Technology, Cyber-Physical Systems (CPS) Programme, and the Smart Manufacturing Leadership Coalition are two of the US government's initiatives to support the development of IoT and CPS. The US government also formed the Advanced Manufacturing Cooperation, a public-private cooperation to encourage the development of advanced manufacturing technologies like IoT. IoT technologies are extensively utilised in Canada in various sectors, including healthcare, agriculture, transportation, and energy. The Strategic Innovation Fund and the Pan-Canadian Artificial Intelligence Strategy are only two of the programmes the Canadian government has created to aid in the development of IoT. The nation's government has also started the Innovation Superclusters Initiative, which aims to promote industry and academic cooperation in the development of new technologies like IoT and supports the growth of several important industries, including advanced manufacturing.

The desire for digital transformation, automation, and data analytics across numerous industries is projected to fuel further growth in IoT usage in America and North America in the upcoming years. However, obstacles must be overcome, including issues with interoperability, data privacy and security, and the requirement for qualified staff to design, develop, and operate IoT systems. IoT adoption is also increasing in Mexico, particularly in the manufacturing industry, a significant engine of the nation's economy. The National Digital Strategy and the National IoT and Big Data Strategy are two activities the Mexican government has started to assist for the growth and acceptance of IoT and CPS technologies. The nation has also built a few innovation centres and technological hubs to encourage the development of new technologies, such as IoT. IoT technologies are also widely used in South America, especially Brazil, Chile, and Argentina. Brazil's government has launched several initiatives to support the development of these technologies, such as the National IoT Plan and the IoT Chamber, and they are being used in a variety of industries, including agriculture, manufacturing, and transportation. The National IoT Policy is one of the government's many attempts to assist the growth of these technologies in Chile, where IoTs are employed in sectors like mining, healthcare, and energy. In Argentina, IoT systems and CPS are utilised in sectors including transportation, healthcare, and agriculture, and the government has started several programmes to assist the growth of these innovations like the National Plan of IoT.

A system of networked objects known as the IoT is equipped with sensors and software that enable it to gather and share data. The IoT can be beneficial for managing water supply networks. By integrating IoT technologies, water supply networks can collect real-time data on various system components, including water quality, flow rate, pressure, and temperature. This information can improve network performance and lower water waste, leakage, and pollution. For instance, sensors positioned throughout the water supply network can find leaks and notify the management system, which can then take action to fix the breaches before they worsen. Similarly, IoT can support real-time water quality monitoring and prompt detection and response to contamination or pollution events. Additionally, IoT can provide remote monitoring and control of the water supply network, enabling operators to operate the network more effectively and make educated decisions in real time. Many of the tasks in managing the water supply network can be automated to lower operating costs, speed up reaction times, and increase system reliability. IoT technology integration into water supply network management can transform how we regulate water resources by improving our systems' effectiveness, resilience, and sustainability. Managing water supply networks can significantly benefit from the fusion of IoT technologies. These benefits include real-time data collection on water quality, flow rate, pressure, temperature, and remote network monitoring and control. IoT can save operational costs, speed up response times, and increase the overall sustainability and reliability of the system by automating operations and empowering operators to make knowledgeable decisions in real time. For instance, Zhang et al. (2016) discovered that IoT technologies can improve network efficiency and water supply systems by reducing wastage. The potential of IoT and CPS for remote monitoring and control of water supply networks is also highlighted by Gavrilovska et al. (2019). While Mohtar et al. (2022) underline their significance for identifying and responding to water contamination events, Khan et al. (2021) explain how these devices can improve water quality management in metropolitan settings.

4.2 IOT COMPONENT AND ARCHITECTURE

The sensing and actuation layer, network layer, and application layer are the three primary parts of the IoT architecture, which are discussed further below:

1. **The sensing and the actuation layer:** Physical objects with sensors and actuators comprise the sensing and actuation layer. Sensors oversee gathering

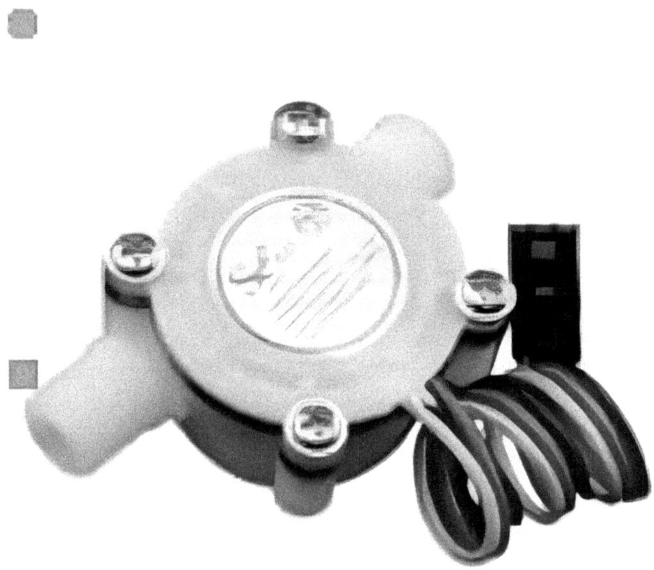

FIGURE 4.2 Water flow sensor used for monitoring and measuring water flow in a system (adapted from Maruthi et al., 2018).

information about various physical and environmental factors, including temperature, humidity, light, and motion. On the other hand, actuators oversee the performance of physical operations like starting a motor or opening a valve. These units, which can operate alone or as a part of larger systems, oversee the production and sending of data to the network layer. For instance, water flow, measuring the volume and rate of water passing through a pipe or channel, is accomplished with a water flow sensor (Figure 4.2). It turns water movement into electrical signals suitable for measurement, data logging, control, or monitoring.

2. **The network layer:** The data gathered by the sensor and actuation layer is processed and transmitted to the application layer by the network layer. This layer allows devices to communicate with one another and with cloud-based services.
3. **The application layer:** It is made up of numerous networking technologies like Wi-Fi, Bluetooth, and cellular networks. Processing data received from the network layer and producing actionable insights and decisions based on that data falls to the application layer. This layer comprises software programmes that let users see, assess, and manage IoT systems and devices. Smart home systems, industrial control systems, and environmental monitoring systems are a few examples of uses.

The IoT reference model and the IoT protocol stack are two examples of models that can represent IoT architecture. The IoT protocol stack outlines each tier's various protocols and technologies. At the same time, the IoT reference model defines the many layers and functionalities of the IoT system (see Figure 4.3).

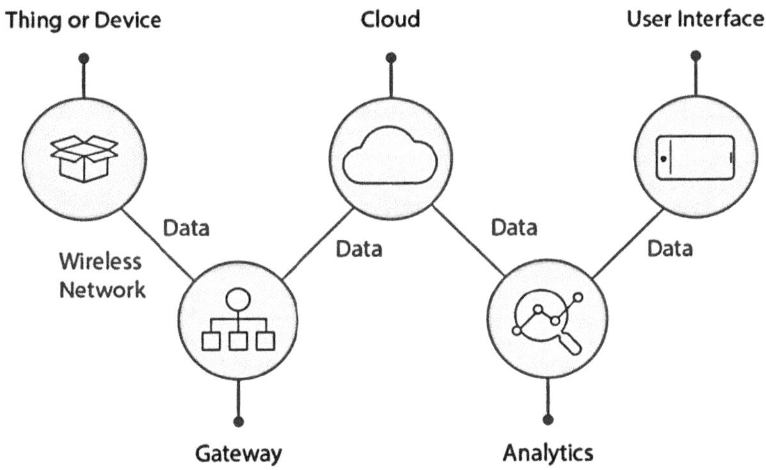

FIGURE 4.3 A simplified Internet of Things (IoT) architecture illustrating the core components, such as sensors, connectivity, data processing, and user interface, working together to enable smart system functionality.

4.2.1 Smart Sensors

Smart sensors are cutting-edge sensing technologies that can process and communicate data to other systems or devices and measure physical and environmental characteristics. Smart sensors have embedded microprocessors, memory, and communication systems that let them interpret and analyse sensor data at a high level. Numerous industries use smart sensors, including healthcare, environmental monitoring, and industrial automation. Smart sensors are employed in industrial automation to track machine performance, find problems, and streamline processes. Smart sensors are employed in environmental monitoring to measure the air, water, and weather quality. Smart sensors are employed in healthcare to track vital signs and spot medical issues. Based on the types of data they gather, such as temperature, pressure, motion, and chemical composition, smart sensors can be categorised. Additionally, they can be categorised according to their communication technologies, such as Wi-Fi, Bluetooth, and Zigbee. Smart sensors can also be developed for smart houses or city systems (Figure 4.4a and b).

Smart sensors' key advantages include their real-time data collection and processing capabilities, minimal power requirements, and interoperability with other systems and devices. Smart sensors have the potential to automate processes, improve operations, and offer insightful data. Smart sensors are sophisticated sensing tools with built-in microprocessors, memory, and networks of communication that let them process and communicate data to other tools or systems. Based on the sort of data they gather, their communication technology, and their design for applications, smart sensors are employed in various applications, and they provide a mechanism to collect data from the real environment and transfer it to other systems and devices. In IoT applications, smart sensors are typically used to track assets, monitor the environment, and report on how well systems and processes work. One of the main

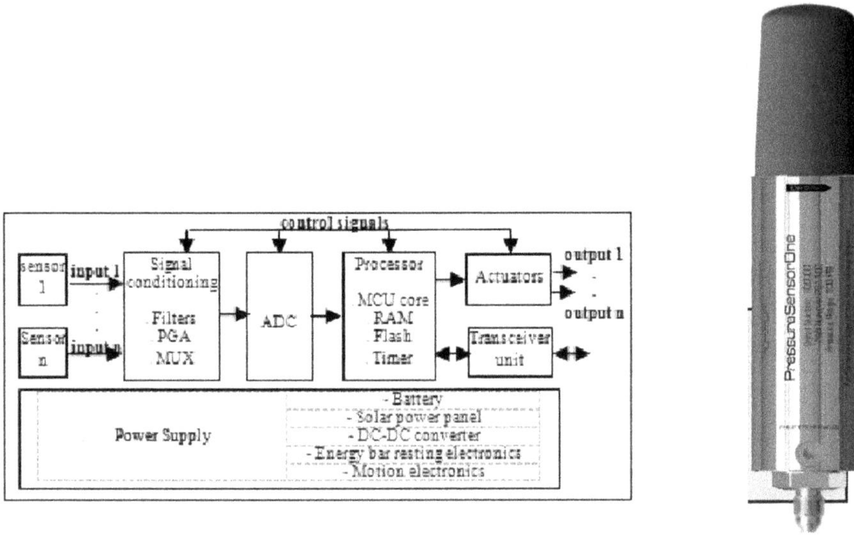

(a) (b)

FIGURE 4.4 Structural and functional components of wireless sensing technology: (a) essential parts of a wireless sensor node, illustrating its modular architecture, and (b) a wireless pressure sensor designed for real-time monitoring applications (adapted from Kayaalpa et al., 2014).

benefits of smart sensors in IoT applications is their ability to communicate with other systems and gadgets. Due to the incorporation of communication technologies like Wi-Fi, Bluetooth, and Zigbee, smart sensors may transmit data to other systems or devices. This makes it possible to analyse and act on the data gathered by smart sensors in real time.

The capacity of smart sensors for sophisticated processing and analysis of sensor data in IoT applications is an additional essential feature. Smart sensors may be configured to conduct intricate computations and analyses on the data they gather, offering insightful information about how well systems and processes work. In IoT applications where devices may be deployed in distant or difficult-to-reach areas, smart sensors can also be engineered to be energy-efficient. Low-power smart sensors require less regular maintenance or replacement because they may run on battery power for extended periods.

4.2.2 Sensor Data Fusion

The IoT uses sensor data fusion technology to merge data from several sensors to create more accurate and thorough information about the monitored environment or system. Sensor data fusion may mitigate errors and provide complete knowledge of the observed system by fusing data from many sensors. Sensor data fusion can be used in IoT applications to increase the accuracy of data gathered from various sensors, each of which may have unique properties or limits. For instance, sensor sampling rates, resolution, or noise characteristics may vary. Sensor data fusion can increase the overall quality and accuracy of the data by integrating data from many sensors. Various algorithms and methods can combine sensor data, including statistical signal processing, machine learning, and artificial intelligence. Such techniques can be applied to data to find patterns, identify anomalies, and extract relevant information. Using a centralised processing system, where all sensor data is transmitted to one location for processing and analysis, is a typical approach for combining sensor data. This method may be helpful in applications where transmission bandwidth is constrained and the sensors are dispersed across a wide area. A decentralised processing system, in which each sensor processes its data and interacts with other sensors to exchange information, is another method for fusing data from several sensors. This strategy may be advantageous in applications with high transmission bandwidth and sensors nearby.

4.2.3 IoT Gateway

An IoT gateway is a device that serves as a link between IoT devices and the Internet or cloud in the context of IoT. IoT devices can connect to the Internet, gather and transmit data, and receive instructions from the cloud using an IoT gateway as a platform. IoT gateways often have several components that allow them to work (Figure 4.5).

These components comprise software and hardware, such as network interfaces, CPUs, and memory, as well as drivers, protocols, and security modules. A variety of sensors and actuators that enable interaction with the real environment may also be included in the gateway. IoT gateways are essential to the IoT ecosystem because they give linked devices a centralised control point. They may carry out tasks like data aggregation, filtering, processing, and analytics, making it possible to operate the IoT network more effectively and efficiently. IoT gateways provide a few critical capabilities, such as the capacity to connect to various IoT devices, support for different communication protocols, and adaptability to various network environments. IoT gateways can also offer security features like encryption and authentication to safeguard the IoT network from threats. In summary, an IoT provides a variety of characteristics, such as data processing, analytics, and security, to enable effective and efficient management of the IoT network.

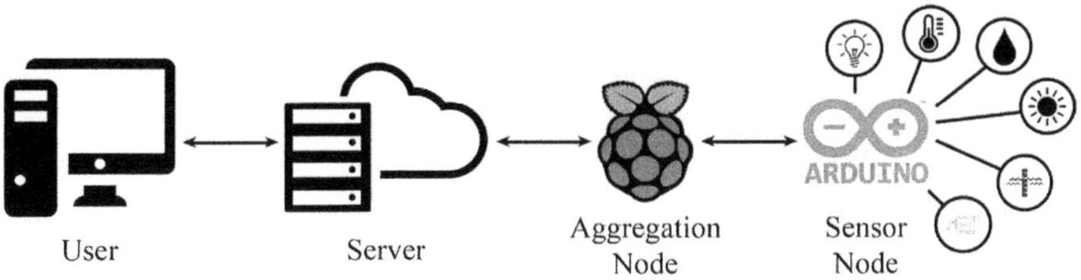

FIGURE 4.5 Architecture of a simple IoT gateway illustrating core components for data acquisition, processing, and transmission in smart water systems (adapted from Glória et al., 2017).

4.2.4 Wireless Communication Networks

A vital part of the IoT ecosystem is wireless communication networks, which allow seamless data transfer between devices, gateways, and cloud-based applications. IoT devices are linked together to the Internet or cloud using wireless communication technology. IoT networks frequently use Wi-Fi, Bluetooth, Zigbee, and cellular networks, among other wireless communication technologies (see Figure 4.6).

4.2.4.1 Wi-Fi

Wi-Fi is a standard wireless technology that connects devices to the Internet and is frequently used in companies and households. High-speed data transfer, little power use, and a wide range are all features of Wi-Fi. Wi-Fi can be used in IoT networks to link IoT devices to a local gateway or the Internet. Bluetooth is another popular wireless communication technology frequently utilised in IoT devices.

4.2.4.2 Bluetooth

Bluetooth is frequently used to link gadgets like smartphones and wearable technology because of its short range. Due to its low power consumption and capacity to connect to several devices at once, Bluetooth low energy is a Bluetooth variation that is frequently utilised in IoT devices.

4.2.4.3 Zigbee

In IoT networks, Zigbee is a low-power wireless communication technology that is frequently utilised for applications like smart lighting and building automation.

FIGURE 4.6 IoT devices are linked together to the Internet or cloud using wireless communication technology. IoT networks frequently use Wi-Fi, Bluetooth, Zigbee, and cellular networks, among other wireless communication technologies.

Zigbee supports numerous devices on a single network and has a range of up to 100 m. Zigbee is frequently utilised in applications for smart buildings and home automation.

4.2.4.4 Cellular Networks

IoT networks use cellular networks like 4G and 5G to connect devices that are situated in far-off or challenging-to-reach places. When compared to other wireless communication technologies, cellular networks are more expensive and offer a wider range and faster data delivery.

4.2.5 Application Management Centre

Application Management Centre (AMC) is a crucial part of the IoT ecosystem. It is a platform for controlling and monitoring IoT applications. The IoT apps, devices, and services are managed and controlled by AMC, which also offers a central site for tracking, evaluating, and reporting on the performance of these applications. Device management, data management, application provisioning, monitoring, and reporting are just a few of the crucial functions the AMC offers for managing IoT applications. The AMC can manage and control the many IoT devices linked to the network, including sensors, gateways, and other connected devices, thanks to the device management capability. The AMC can ensure that devices are correctly configured and are effectively connecting to the network, owing to this capability.

The AMC's data management capability offers a framework for managing the enormous amounts of data that IoT devices produce. This feature enables the construction of dashboards and reports that offer real-time insights into the functionality of the IoT applications, as well as the storage, processing, and analysis of the data by the AMC. Administrators can deploy and manage IoT apps, including software updates and security patches, using the AMC's application provisioning function. This functionality makes sure that the IoT applications are safe and secure and are guarded against threats from the Internet. Foremost, the AMC's monitoring and reporting capability offers real-time visibility into the performance of IoT applications, devices, and services. With the help of this functionality, administrators may immediately identify and fix problems, and they can gain insights into how to improve the IoT ecosystem's performance.

4.3 APPLICATIONS OF IOT TO WATER SUPPLY NETWORK MANAGEMENT

The management of water supply networks can benefit from a variety of IoT applications mentioned below.

4.3.1 Smart Water Meters

IoT-enabled water meters can remotely shut off the water supply in case of emergencies or unpaid bills, monitor water usage in real time, find leaks, and identify possible problems. This can increase operating effectiveness and decrease water loss.

4.3.2 Water Quality Monitoring

IoT sensors can monitor water quality indicators in real time, including pH, turbidity, temperature, and dissolved oxygen levels. This data can be utilised to identify contamination occurrences and send out real-time notifications, allowing for quick responses.

4.3.3 Predictive Maintenance

With IoT-enabled sensors, water network infrastructure, including pipes, pumps, and valves, can be continuously monitored for efficiency. This can enable predictive maintenance and decrease downtime by detecting possible faults before they develop into serious ones.

4.3.4 Flood Monitoring and Warning Systems

IoT sensors can monitor water levels in rivers, reservoirs, and other bodies of water in real time. To facilitate prompt evacuation and damage alleviation, this can assist in identifying future flooding disasters and issuing early warnings to local authorities and citizens.

4.3.5 Water Resource Management

IoT technology can continuously monitor the water levels in reservoirs, rivers, and other water sources. The distribution and allocation of water resources can be optimised using this information, which may also be used to anticipate future water shortages.

4.3.6 Tracking Assets

IoT sensors can keep tabs on the whereabouts, condition, and performance of water network assets like pipes, valves, and pumps. This can aid in maximising asset use, enhancing maintenance planning, and decreasing the likelihood of asset failure.

4.3.7 Energy Management

Water treatment facilities, pumping stations, and other facilities can track their energy usage using IoT-enabled sensors. This can assist in locating potential for energy efficiency and operating cost reduction.

4.3.8 Engagement of the Customers

IoT-enabled smart meters can give customers real-time feedback on their water usage, allowing them to track and modify their consumption patterns. This could encourage water conservation and help with peak demand reduction.

4.3.9 Leak Detection

Real-time monitoring of water pressure and flow rates with IoT sensors enables the early identification of leaks in the water supply network. This can lessen the danger of flood damage to infrastructure and property while also reducing water loss.

4.3.10 Demand Forecasting

IoT technology can be utilised for data collection, analysis, and demand forecasting on water consumption trends. Using this data will improve water supply planning and lessen the possibility of water shortages.

4.3.11 Irrigation Management

By using IoT sensors to track meteorological data and soil moisture levels, farmers can better plan their irrigation systems and use less water.

4.3.12 Optimisation of Water Treatment

Optimisation of the water treatment process with IoT sensors includes monitoring chemical dosage, flow rates, and other elements. Both operating costs and treatment effectiveness may benefit from this.

4.3.13 Remote Monitoring and Control

IoT technology allows for remote monitoring and control of water supply infrastructure, eliminating the need for on-site inspections and facilitating quicker problem-solving.

4.3.14 Water Conservation

Customers can measure their usage and find chances for water saving by using IoT devices to monitor water usage in real time. Doing so may encourage sustainable water management techniques and lower total water consumption.

These applications show the breadth of ways, IoT technology may be applied to enhance water supply network management, from streamlining treatment procedures to controlling irrigation and lowering the danger of flooding incidents. It shows off the adaptability of IoT technology in managing water supply networks and its potential to have a significant impact on cost savings, resource optimisation, and customer service.

4.4 CHALLENGES AND PROGRESS

The introduction of IoT in water supply network management faces various hurdles, as with any newly developed technology. The following are some of the primary obstacles to IoT water supply network management implementation:

1. **Data Management:** IoT sensors produce enormous data, which can be challenging to manage and analyse. To solve this problem, new data management and analysis tools are being created to speed up processing massive amounts of data.
2. **Security:** Cyberattacks on IoT networks can jeopardise the integrity and security of the water supply network. New security protocols and procedures are being created to safeguard IoT networks to handle this challenge.

3. **Cost:** For many water companies, the expense of integrating IoT technology might be a major roadblock. Greater financing and effort are required to assist in the creation and application of IoT solutions to address this challenge.
4. **Interoperability:** It may be challenging to combine several systems into a single network since IoT devices and systems from various suppliers may not be compatible. Industry standards and protocols are being created to assure compatibility between various IoT systems to overcome this difficulty.

Despite these obstacles, the application of IoT in water supply network management has advanced significantly. IoT technology has seen an increase in adoption in the management of water supply networks, with many water utilities deploying IoT solutions to increase network effectiveness and cut costs. The development of IoT sensors, platforms, and apps for managing water supply networks has seen a lot of innovation, with new technologies being created to address problems facing the sector. Also, to develop and implement IoT solutions for water supply network management, there has been an increase in collaboration between industry stakeholders, including water utilities, technology suppliers, and government agencies. Furthermore, more people are becoming aware of the potential advantages of IoT technology for managing water supply networks, which is encouraging more investment and support for the creation and adoption of IoT solutions. Although there are still issues to be resolved, the development of IoT in the management of water supply networks through adoption, innovation, collaboration, and awareness points to the technology's great potential to enhance the effectiveness, sustainability, and dependability of water supply networks in the years to come.

4.5 CHAPTER SUMMARY AND CONCLUSION

Implementing IoT technology has been applied to various aspects of the water supply system, aligning with the objectives of the Smart Water Network Management effort. This initiative seeks to introduce a systematic approach to enhance water supply systems' operational efficiency and monitoring capabilities. The chapter presents the uses of IoT in monitoring the water supply network, with a specific focus on its utilisation for water quality assessment and leakage detection. The water management system, which is based on IoT, employs a range of interconnected devices such as sensors, controllers, and meters. These devices are integrated with mobile and online applications, as well as tools for data processing and analysis. These platforms will facilitate effective management of water supply, monitoring freshwater quality, detection of pollution, and other related activities.

REFERENCES

Bandyopadhyay, D., & Sen, J. (2015). Internet of Things applications for water quality assessment. In *Internet of Things and Big Data Technologies for Next Generation Healthcare* (pp. 31–45). Springer. https://doi.org/10.1007/978-3-319-19890-3_3

Gavrilovska, L., Chatzimisios, P., & Gjorgjevikj, D. (2019). Cyber-physical systems in smart water management: A survey. *Journal of Network and Computer Applications*, *136*, 79–93.

Glória, A., Cercas, F., & Souto, N. (2017). Design and implementation of an IoT gateway to create smart environments. *Procedia Computer Science*, *109*, 568–575.

Kagermann, H., Wahlster, W., & Helbig, J. (2013). *Recommendations for Implementing the Strategic Initiative Industrie 4.0: Final Report of the Industrie 4.0 Working Group*. Acatech – National Academy of Science and Engineering.

Kayaalpa, F., Zenginb, A., Karac, R., & Buyukguclua, M. A. (2014). Water powered sensors in water distribution systems. *Turkish Journal of Engineering*, *1*, 24–31.

Khan, M. A., Alam, M. M., Gupta, R., & Kim, D. H. (2021). Internet of Things and cyber-physical systems for improved water quality management in urban areas: A review. *Journal of Cleaner Production*, *297*, 126690. https://doi.org/10.1016/j.jclepro.2021.126690

Kumar, A., Gupta, A., Singh, A., & Nigam, M. J. (2017). IoT-based smart water management system for efficient irrigation. *Wireless Personal Communications*, *97*(4), 5637–5654.

Maruthi, H. V., Priya, L., Lavanya, A. R., Manideep, M., & Jayannavar, L. (2018, May). An IoT-based water supply monitoring and controlling system. In *International Journal of Advanced Research in Computer Science (ISSN: 0976-5697) Conference Paper: Third National Conference on "Advances in Computing and Information Technology*, 9(Special Issue).

Mohtar, R. H., Saeed, A. O., & Syed-Hassan, S. S. (2022). A review of the potential use of cyber-physical systems and the Internet of Things in water security. *Journal of Cleaner Production*, *336*, 127194. https://doi.org/10.1016/j.jclepro.2021.127194

Zhang, Y., Xiao, Y., Li, H., Li, H., & Li, G. (2016). A cyber-physical system approach to water quality monitoring in distribution systems. *Water Research*, *102*, 327–337. https://doi.org/10.1016/j.watres.2016.06.031.

5 CPS in the Water Supply Network Management

5.1 INTRODUCTION

Cyber-physical systems (CPS) can be used to integrate and optimise the various water supply network components, such as sensors, pumps, valves, and control systems, in managing the water supply network. CPS collects data on water quality, pressure, flow, and other important characteristics while monitoring the water supply network. Pumps, valves, and other network components can then be optimised using this data to guarantee that water is delivered effectively and efficiently. Using data analytics and machine learning techniques to find patterns and abnormalities in sensor data, CPS can be used to discover and diagnose issues in the water supply network before they arise. This may aid in lowering downtime and enhancing network dependability.

A CPS can also be used to track and regulate the quality of the water flowing through the water supply network. It does so by utilising sensors and data analytics to identify changes in the water's quality and adjusting the way treatment systems are run to retain water quality at a high level. Using data analytics and control algorithms, CPS can optimise the energy consumption of the water supply network, adjusting the operation of pumps, valves, and other network components to reduce energy use while ensuring a dependable water supply. While CPS has several advantages for managing water supply networks, some obstacles must also be overcome. These difficulties require the development of cutting-edge sensor technologies, dependable and secure communication networks, and efficient data management and analytics systems.

5.2 CPS COMPONENTS AND ARCHITECTURE

CPS can be generally separated into physical, computational, and communication components.

1. **Physical Components:** This category includes CPS's tools to monitor and regulate physical processes. Physical components in managing the water supply network may consist of pumps, valves, and other apparatus used to move and treat water, and sensors that assess water quality, pressure, and flow.
2. **Computational Components:** These programmes and formulas process and examine the information gathered by the physical components. Machine learning algorithms used to analyse sensor data and find patterns and anomalies and control algorithms used to modify the operation of pumps and valves to optimise the performance of the water supply network

DOI: 10.1201/9781003633839-5

are examples of computational components that could be used in water supply network management.

3. **Communication Components:** The CPS uses networks and protocols to communicate information between its physical and computational parts. Communication components in water supply network management may include wireless networks that transport data from sensors to a central control system and communication protocols that allow communication between various network components.

A CPS's architecture (Figure 5.1) typically comprises several levels, each with its own set of components and functionalities. The layers could consist of:

1. **Sensing Layer:** The physical elements utilised to gather information about the physical process that the CPS oversees, making up this layer. Water quality, pressure, and flow sensors may be a part of managing the water supply network.
2. **Network Layer:** This communication tool connects the CPS's physical and computational components. This could include wireless networks that send data from sensors to a centralised control system in the context of water supply network management.

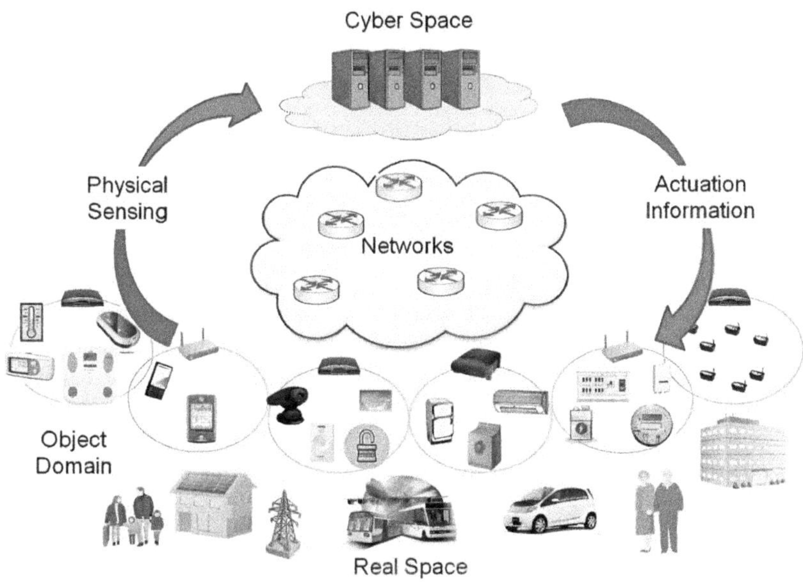

FIGURE 5.1 Cyber-physical system (CPS) architecture showing how smart objects in the physical/object domain sense conditions in the real space, transmit data through communication networks to cyber space for processing and decision-making, and then send actuation information back to physical devices and infrastructure to close the control loop.

Source: Raisin et al. (2020).

3. **Data Processing Layer:** The tools for processing and analysing the information gathered by the sensing layer are found in this layer. This could involve machine learning algorithms that evaluate sensor data and find trends and abnormalities in managing the water supply network.
4. **Control Layer:** This layer's computational elements control the computational parts of the CPS's physical components. This could include control algorithms used to modify the functioning of pumps and valves in managing the water supply network to maximise the efficiency of the water supply network.
5. **Application Layer:** The software programmes used to visualise and handle the data gathered and processed by the CPS are included in this layer. This could refer to dashboards and other tools that water supply managers use to monitor and manage the network in the context of water supply network management.

The physical system, sensor system, actuation system, communication network, and cyber and decision support system are all represented in Figure 5.1. They are comprised of different parts of a CPS. The real-world system being observed and managed is called the physical system, and the sensing system collects data from the physical system. Based on this information, the actuation system subsequently controls the physical system. The cyber and decision support system processes and analyses the data to make decisions and offer feedback to the other components. The communication network facilitates a data flow between these components.

5.2.1 CPS Sensing System

Sensing systems are employed in water supply network management to gather data on water flow, pressure, quality, and other pertinent factors (Dziewierz et al., 2017; Li et al., 2020). The control system receives the data after that for analysis and decision-making. Different types of sensors, including flow meters, pressure sensors, temperature sensors, water quality monitors, and humidity sensors, can make up a CPS's sensing system. These sensors can be dispersed over the water supply network to track various metrics in various places. The control system can receive the data gathered by these sensors via wired or wireless communication methods like Wi-Fi, LoRaWAN, or Zigbee. The distance between the sensor and the control system, the volume of data to be transferred, and the sensor's power requirements affect the communication method. The sensors in a CPS need to be calibrated frequently and maintained adequately to guarantee dependable and accurate sensing. Any sensor failure or malfunction might produce inaccurate data, thereby influencing the control system's decisions. Most of the time, sensing systems in a CPS use a network of sensors dispersed across the water supply network to track various variables, including water flow, pressure, temperature, water quality, and more. Sensors can be interconnected to create a real-time network that can relay data to a central control system (Adedeji et al., 2019). The water supply network can use these sensors in many places, including reservoirs, pumping stations, treatment facilities, distribution pipelines, and end-user connections. Depending on the location and the needs of the sensor, they can be powered by batteries, solar panels, or cable power. Ensuring that the data collected by the sensors are accurate and trustworthy is one of the main problems in deploying sensing systems in water supply networks. The sensors must be calibrated and maintained regularly to keep operating correctly. Any variations or errors in the sensor readings could produce inaccurate data and harm the CPS's functionality. A CPS's sensing system must also be built to withstand outside influences like severe weather, vandalism, and other forms of interference. To ensure that data can be gathered even if specific sensors malfunction, this may entail employing ruggedised sensors that can tolerate severe environments or building redundant sensor networks.

5.2.2 CPS Actuation System

The component that performs actions because of the analysis of data gathered by the sensing system is known as the actuation system in the CPS for water supply network management (Wang et al., 2019). It contains automation and control technologies that make it possible to optimise and modify water supply processes in real time. According to Huang et al. (2018) and Liu et al. (2018), the actuation system gets data from the sensing system, processes it using sophisticated algorithms, and then initiates the proper actions based on the outcomes. It might, for instance, modify the pressure or flow rate of the water in response to shifting demand, find and fix leaks or other issues, and regulate pumps and other equipment to save energy use. According to Zeng et al. (2017), the actuation system frequently consists of different kinds of controllers, including programmable logic controllers, distributed control systems, and supervisory control and data acquisition systems. These controllers frequently use wired or wireless communication technologies to connect to sensors and other devices.

5.2.3 CPS Communication Networks

To provide a smooth connection between physical systems and computational entities, CPS rely on communication networks. Communication networks are employed in the administration of water supply networks to gather data from sensors, transfer it to a central control system, and provide control instructions to actuators. To facilitate real-time data interchange and decision-making across system components, Wang et al. (2019) claim that communication networks are a crucial part of the CPS for managing water supply networks. The total performance of the CPS depends on how well the communication network functions, as delays, packet losses, and other network problems can significantly reduce performance or cause the system to fail. CPS has suggested and utilised different communication network technologies to ensure timely and reliable communication. Wired networks, wireless networks, and hybrid networks that integrate wired and wireless technologies are some of CPSs' more well-liked communication network technologies.

Physical cables are used in wired networks to deliver data between devices. Ethernet and field bus networks are two wired networks frequently utilised in CPS.

Ethernet's high bandwidth and low latency make it a popular choice for industrial automation and CPS. Contrarily, field bus networks are used in the process automation sector to manage field equipment like sensors and actuators.

On the other hand, wireless networks use radio waves to transfer data between devices. Wireless networks like Wi-Fi, Bluetooth, Zigbee, and cellular networks are standard wireless networks used in CPS. While Zigbee is frequently employed in low-power, low-data-rate applications, Wi-Fi and Bluetooth are preferred for local area communication. On the other hand, cellular networks are utilised in CPS for long-distance communication, particularly in isolated areas. Xie et al. (2019) studied wireless sensor networks (WSNs) in CPS communication networks and emphasised the advantages of WSNs for dependable and effective data transmission in water supply network management. In their study, Zhang et al. (2018) presented a CPS-based communication network for managing water supply networks. This network combines various communication technologies to enable real-time monitoring and control of system components. Dong et al. (2019) examined the difficulties and emphasised the significance of secure and trustworthy communication networks in CPS for managing the water supply network. Lu et al. (2018) studied the role of cloud computing in CPS communication networks for managing water supply networks. They highlighted the advantages of cloud-based data processing and analysis for enhancing system performance.

Hybrid networks mix wired and wireless networks to benefit from each type's advantages. For instance, a hybrid network can use wireless networks for remote communication or when mobility is necessary, while wired networks are used for high-bandwidth, low-latency applications. Numerous strategies have been developed and implemented to increase the dependability and performance of communication networks in CPS. These include security measures, Quality of Service guarantees, fault tolerance, and redundancy. Redundancy includes duplicating crucial network components to ensure the system can continue functioning even if certain aspects collapse. Designing the network for fault tolerance entails making it resilient to errors and failures without degrading its functionality. To ensure timely delivery, quality of service guarantees must be made to prioritise vital data over non-critical data. Security measures against hacker attacks, data theft, and other intrusions safeguard the network.

5.2.4 CPS Cyber and Decision Support Systems

In managing water supply networks, cyber and decision support systems are essential to CPS. By offering real-time monitoring, data analysis, and support for decision-making, they contribute to ensuring the efficient and successful operation of the system. Various hardware and software parts that offer real-time monitoring and control of the physical system comprise the CPS's cyber system portion. These consist of user interfaces, communication networks, data storage and processing systems, actuators, sensors, and controllers. Analytical tools and algorithms use data from the physical system in the decision-support system component to support decision-making processes. For instance, in a network management system for a water supply, the cyber system component of the CPS may comprise sensors that check the quantity and quality of the water, as well as pumps and valves that regulate the water's flow. Communication networks may send data between physical and cyber systems using wireless and wired technologies, including Wi-Fi, Zigbee, and LoRaWAN. The decision support system part of the CPS can then use machine learning algorithms and other analytical tools to process and evaluate the data gathered from the physical system to discover abnormalities, forecast system breakdowns, and enhance system performance. The efficiency of CPS cyber and decision support systems in managing water supply networks has been shown in numerous studies. For instance, Jin et al.'s study in 2021 created a CPS-based decision support system for water supply networks that included simulation and optimisation methods to determine the best places for water quality sensors. A data-driven problem detection and diagnostic system, as well as a multi-objective optimisation algorithm for system operation, were incorporated in another study conducted by Chen et al. (2019), which proposed a CPS-based framework for managing water supply networks.

5.3 FUNCTIONAL ATTRIBUTES OF THE CPS

CPS are distinguished by their close integration with the digital world and their capacity to perceive, process, and act on the physical world. CPS are employed in numerous fields, such as the administration of water supply networks, transportation systems, energy systems, and healthcare systems.

Three broad categories can be used to group the functional characteristics of CPS:

1. Control,
2. Communication, and
3. Computation.

Control refers to CPS's capacity to control the physical world via actuators and sensors. The ability of the CPS to communicate with its many parts, including its sensors, actuators, and decision-making algorithms, is called communication. The term "computation" describes the CPS's capacity to process the data obtained from sensors and other system elements and use it to decide what to do and how to do it.

5.3.1 Control

One of the most crucial characteristics of CPS is control, since it allows the system to interact with the real environment. Actuators, which include valves, pumps, and motors, can alter the state of the physical universe. On the other hand, sensors give the system information on the state of the physical world right now. The CPS's control algorithms must be built to guarantee that the system reacts appropriately to modifications in the physical world.

5.3.2 Communication

Another crucial characteristic of CPS is communication, which permits information flow between the system's many components. Communication between sensors, actuators, and decision-making algorithms frequently happens in the CPS. The CPS's communication protocols must be created to guarantee that the system can function dependably in various settings and circumstances.

5.3.3 Computation

Another crucial characteristic of the CPS is computation, which enables the system to process data from sensors and other system components. The system can make judgements and conduct actions based on the status of the physical world, thanks to the data processing capabilities of the CPS. The computing capabilities of the CPS must be designed such that the system can function in real time and under pressure.

5.4 APPLICATION OF CPS TO WATER SUPPLY NETWORK MANAGEMENT

Water quality monitoring, leak detection, pressure management, and energy efficiency are just a few facets of water supply network management. The CPS has been applied in many forms.

5.4.1 Water Quality Monitoring

A network of sensors and actuators can be utilised with the CPS to monitor the water quality continuously. For instance, Adedeji & Hamam (2020) developed a CPS-based water quality monitoring system that combines physical and chemical sensors to find contaminants in the water supply network and notify the authorities for quick action. This makes it possible to quickly detect and react to any alterations in the quality of the water, such as contamination, as illustrated in Figure 5.2.

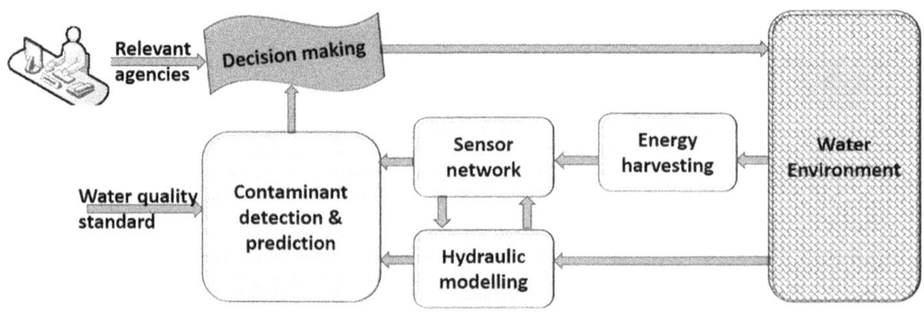

FIGURE 5.2 Water quality monitoring framework showing how sensor networks deployed in the water environment, supported by energy-harvesting units and hydraulic modelling, generate data for contaminant detection and prediction against predefined water quality standards. The resulting information is fed to relevant agencies to support informed decision-making and timely management actions for maintaining safe water resources (Adedeji & Hamam, 2020).

5.4.2 Leak Detection and Repair

Real-time monitoring and leak detection in the water delivery network is also possible with CPS. Liu et al. (2017) created a CPS-based leak detection system that uses acoustic sensors to find pipeline leaks. This can lessen the chance of infrastructure damage and assist in identifying and preventing water waste. The technology can then warn maintenance workers to make repairs and automatically turn off the water supply.

5.4.3 Pressure Management

CPS can manage water pressure in the supply network to maintain optimum performance and energy effectiveness. For instance, Wang and Wang (2017) created a CPS-based pressure management system that regulates pipeline pressure using a network of sensors and actuators.

5.4.4 Energy Efficiency or Optimisation

Pumps and other equipment in the water delivery network can optimise their energy consumption using CPS. For instance, Chen et al. (2018) created a CPS-based energy management system that uses a mix of sensors and actuators to enhance the performance of the pipeline's pumps and valves. The system can alter the operations of the equipment to decrease energy usage by interpreting the data gathered by the sensors. Additionally, it can aid in lowering energy costs and enhancing the network's general sustainability.

5.4.5 Predictive Maintenance

The water delivery network's infrastructure and equipment can be monitored using the CPS. The system can forecast when maintenance or repair work is necessary by evaluating the data gathered by the sensors, preventing equipment failure, and reducing downtime.

5.4.6 Demand Forecasting and Emergency Response

Based on historical usage patterns and other pertinent variables, CPS can be used to forecast the demand for water in the supply network. This can aid in better water supply network planning and management, as well as aid utilities in more effective water distribution management and waste reduction. In addition, CPS can be used to react swiftly to catastrophes brought on by natural disasters like floods and droughts. Technology can give authorities access to real-time data on the state of the water supply network, enabling them to take the required steps to protect the public's safety and well-being.

5.4.7 Water Conservation and Smart Metering

By giving customers up-to-date data on their water usage, CPS can be used to encourage water conservation. Customers may be able to modify their behaviour and use less water. The water supply network can employ a CPS to develop smart metering systems. This can aid in improved invoicing and revenue collection management and give customers access to real-time water usage data.

5.4.8 Water Resource and Asset Management

Water resources in the supply network can be monitored and managed with the help of CPS. Technology can assist in better managing water resources and ensuring their sustainability by accessing data on availability and usage trends. Pumps, valves, and other equipment utilised in the water delivery network may all be managed and tracked using the CPS. The system can forecast when maintenance or replacement work is necessary by continuously monitoring the condition of these assets, which helps to ensure their dependability and lifetime. This enables better planning of maintenance and upgrades and ensures that resources are spent effectively.

5.4.9 Water Loss Management and Customer Engagement

The CPS can track water loss in the distribution network and pinpoint places that can be improved to cut losses. Customers can receive real-time information on their water usage from the CPS, motivating them to use water more wisely.

5.5 CHALLENGES AND PROGRESS

CPS in water supply network management involve several obstacles, as with any newly developed technology. The interoperability of various CPS components and subsystems, which may use multiple communication protocols and data formats, represents one of the main problems. This might make it challenging to synchronise and handle data from several sources. The security of CPS systems, which could be open to hacker assaults and illegal access, is another issue. This is especially important in water supply networks, which are essential pieces of infrastructure that must be safeguarded from online dangers. Additionally, CPS systems need to be exceptionally highly reliable and fault-tolerant because even tiny errors can have unfavourable adverse effects on the water supply network. Robust design, testing, and maintenance procedures—which can be costly and time-consuming—are necessary for this.

Despite these obstacles, CPS in water supply network management have made great strides. Recent studies have demonstrated that the CPS can decrease operational costs and environmental impact while increasing the water supply network's efficiency, reliability, and resilience. By dynamically modifying pump and valve settings in response to real-time data on water demand, pressure, and quality, for instance, CPS can be used to optimise water distribution. This can enhance the network's performance by lowering energy use and water loss. Additionally, the CPS can facilitate early leak and other network anomaly detection, allowing for prompt maintenance and action. This can help reduce expensive water losses and water supply disruptions. By providing real-time data and analytics on network performance, which can help with strategic planning and resource allocation, CPS can also support decision-making in managing the water supply. Implementing CPS to water supply network management has significant potential for enhancing the sustainability and resilience of our water resources. However, it also comes with many technical, operational, and security concerns that must be carefully considered.

5.6 CHAPTER SUMMARY AND CONCLUSION

CPS have immense promises for managing the water delivery network. A CPS merges physical and cyber systems to create an intelligent system that can gather and process data in real time for decision-making, including sensing, actuation, and communication systems. A CPS also uses sophisticated decision support systems to provide the best decision-making based on real-time data. Monitoring water quality, finding leaks, performing preventive maintenance, and water conservation are some CPS applications in managing the water supply network. To fully achieve the potential of these technologies, specific issues, including data security, privacy, interoperability, and scalability, must be resolved. The advancement of CPS technologies, such as smart sensors, data fusion methods, and sophisticated decision support systems, is spurring the creation of cutting-edge approaches to managing water supply networks. These technologies will continue to revolutionise the water supply network management and deliver dependable and effective services with additional research and development.

REFERENCES

Adedeji, K. B., & Hamam, Y. (2020). Cyber-physical systems for water supply network management: Basics, challenges, and roadmap. *Sustainability*, *12*(22), 9555.

Adedeji, K. B., Nwulu, N. I., & Clinton, A. (2019, September). IoT-based smart water network management: Challenges and future trend. In *2019 IEEE AFRICON* (pp. 1–6). IEEE.

Chen, G., Li, L., Wang, Y., Li, X., & Yang, L. (2019). A cyber-physical system-based framework for water supply network management. *Journal of Hydroinformatics*, *21*(2), 340–353.

Chen, W., Guan, X., Jia, Q.-S., & Song, Z. (2018). Cyber-physical system for energy management in water supply networks. *IEEE Transactions on Industrial Informatics*, *14*(1), 159–168.

Dong, X., Yu, Y., Chen, H., & Li, H. (2019). Research on communication security of the cyber-physical system in the water supply network. In *Proceedings of the 3rd International Conference on Control Science and Systems Engineering* (pp. 264–270). IEEE.

Dziewierz, D., Swiatek, P., Gburzynski, J. P., & Pieniazek, M. T. (2017). Development of a smart water supply system based on cyber-physical systems. *Sensors*, *17*(2), 1–23. https://doi.org/10.3390/s17020370

Huang, G., Wu, S., Chen, Z., & Chen, J. (2018). Design and implementation of a cyber-physical system for water supply network management. *IEEE Access*, *6*, 759–771.

Jin, X., Chen, Z., Ma, J., & Lu, J. (2021). A cyber-physical system-based decision support system for water supply networks. *Journal of Cleaner Production*, *280*, 124323.

Li, J., Li, Y., Gao, L., Li, J., & Song, H. (2020). Cyber-physical system for intelligent water supply networks: A review. *Journal of Cleaner Production*, *247*, 119131.

Liu, X., Wei, Z., Gao, Z., & Tan, Y. (2017). Cyber-physical system for pipeline leak detection based on acoustic signal analysis. *IEEE Transactions on Instrumentation and Measurement*, *66*(5), 793–801.

Liu, Y., Guo, J., Liu, Y., Wang, J., & Huang, S. (2018). Cyber-physical system based water supply network optimisation: A review. *Journal of Cleaner Production*, *195*, 119–129.

Lu, J., Wang, Y., Cao, J., & Shao, Z. (2018). A cyber-physical system for water supply network based on cloud computing. *IEEE Access*, *6*, 12152–12159.

Raisin, S. N., Jamaludin, J., Rahalim, F. M., Mohamad, F. A. J., & Naeem, B. (2020). Cyber-physical system (CPS) application-a review. *REKA ELKOMIKA: Journal Pengabdian Kepada Masyarakat*, *1*(2), 52–65.

Wang, L., & Wang, Y. (2017). Cyber-physical systems: A review of research challenges and directions. *Journal of Systems Science and Complexity*, *30*(1), 43–62.

Wang, Y., Fang, Y., Li, J., Li, J., & Li, Y. (2019). Cyber-physical system for water supply network management: A review. *IEEE Access*, *7*, 141110–141127.

Xie, H., Li, J., Li, Y., Huang, Z., & Fang, Y. (2019). Application of wireless sensor network in cyber-physical system for water supply network. In *Proceedings of the 3rd International Conference on Advances in Energy and Environment Research* (pp. 60–66). Atlantis Press.

Zeng, F., Huang, W., Li, W., & Jiang, Y. (2017). Design and implementation of a cyber-physical system for the water supply network. *Journal of Water Resources Planning and Management*, *143*(5), 04017012.

Zhang, J., Yu, Z., & Shao, W. (2018). Cyber-physical system for water supply network management: A review. *Journal of Water Supply: Research and Technology-Aqua*, *67*(7), 610–622.

6 Intelligent Computing in Water Supply Network Management

6.1 INTRODUCTION

The definition of intelligent computing has yet to be universally captured to an agreed point. However, it can be viewed as an amalgamation of artificial intelligence (AI) and computing technology, with significant improvements traceable to AI development. This definition, however, confines intelligent computing to the field of AI, abhorring the inherent limitations of AI and its indispensable role among humans, machines, and interaction with the environment. Intelligent computing can also be viewed as computational intelligence that imitates biological or human intelligence to initiate optimal algorithms to solve problems. The definition still needs to consider the importance of computing architecture and Internet of Things (IoT) in intelligent computing.

Another definition can be framed for the fusion of three fundamental spaces: human society space, physical space, and information space. The definition addresses complex scientific and societal problems that may arise from the combination of spaces. Hence, intelligent computing is a field comprising a theory (or theoretical method), an architectural system, and a technical power in the digital era, fostering global interaction. The main objective is to create optimised computational tasks by minimising costs based on specific requirements, suiting computational power, employing optimal algorithms, and achieving optimal outcomes. This restructured explanation of intelligent computing comes from escalating computational demands that result from the human societal, physical world, and information space convergence. It prioritises human-centred approaches, striving for high computing capability, energy efficiency, intelligence, and security. The main goal is to provide autonomous, efficient, reliable, secure, transparent, and universal computing services to support large-scale and intricate computational risk.

Intelligent computing and automation play a crucial role in managing water supply networks. These technologies leverage advanced algorithms, data analysis, and control systems to optimise the operation and maintenance of water distribution systems. Water supply network management is critical in ensuring clean water's efficient and sustainable delivery to communities. With the growing complexities and challenges water distribution systems face, there is an increasing need to leverage advanced technologies to enhance their performance. Intelligent computing has emerged as a transformative tool in managing water supply networks, offering novel approaches for real-time monitoring, predictive analytics, fault detection, and

DOI: 10.1201/9781003633839-6

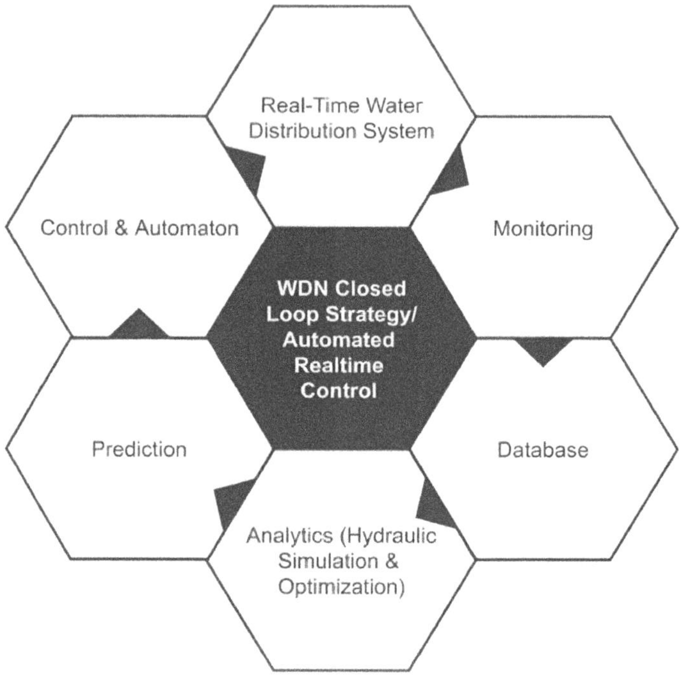

FIGURE 6.1 Closed-loop strategy for a water supply system, illustrating how automated real-time control of the water distribution network (WDN) is achieved through the interaction of continuous monitoring, centralised databases, hydraulic simulation and optimisation analytics, predictive tools, and control and automation modules, all integrated with the real-time water distribution system (Valeyudhan et al., 2022).

decision support. This chapter explores the applications and benefits of intelligent computing in water supply network management, backed by empirical evidence and research studies. Traditional water supply network management approaches often involve manual interventions and reactive strategies. However, with advancements in computing power, data analytics, and control systems, intelligent computing has revolutionised the field. This technology enables water operators to proactively monitor and control various network parameters, optimise operational efficiency, reduce energy consumption, and minimise water losses (see Figure 6.1) for the strategy used in the real-time water distribution network. By integrating real-time data, predictive analytics, and decision support systems (DSS), intelligent computing empowers water supply networks to operate in a more informed and adaptive manner.

Intelligent computing techniques like machine learning and AI enable predictive water supply network management analytics. These algorithms analyse historical data to identify patterns and correlations, allowing operators to forecast future demands, detect potential failures, and optimise maintenance schedules. Fault detection algorithms can automatically identify abnormalities in the system, such as leaks or pipe bursts, enabling a timely response and reducing the risk of water loss and infrastructure damage (Liu et al., 2021). Intelligent computing techniques contribute to developing DSS for water supply network management. These systems integrate real-time data, historical records, and mathematical models to provide operators with insights and recommendations. DSS can assist in optimal pump scheduling, demand forecasting, water allocation, and resource management. By leveraging intelligent computing, DSS enhance the efficiency and resilience of water supply networks (Maier et al., 2014). Intelligent computing technologies are crucial for developing smart water grids. These grids integrate sensors, actuators, and communication networks to enable real-time data exchange and control in water supply networks. Smart water grids leverage intelligent computing algorithms to optimise water distribution, detect and respond to leaks or failures, and give customers timely information about their water usage. These systems enhance network efficiency, reduce water losses, and improve water supply management (Lian et al., 2019).

This system (Figure 6.2) proposes an interface where the digital signal processor (DSP) interacts with components like the image sensor, Global System for Mobile Communications (GSM) modem, and solenoid valve. A mechanical meter equipped with an image captures the reading on the meter upon receiving an impulse of stimulation from the central server during the billing generation cycle. The DSP captures and transforms the order to extract the meter reading, which is returned to the central server for analysis and billing. Software (automatic billing generation) designed to generate bills for the meter reading received is sent to the consumer through their mobile phones. In addition, the software is used by the system to analyse the reading on the meter, detecting any leakages or potential leakages in any zone. Any payment will be updated on the server, and any household that does not pay its water utility bill will have its network disconnected from the central server. A message will then be sent to each meter to stop the water supply, and the DSP activates the solenoid valves.

Groundwater is crucial in meeting the water needs of households, agriculture, and commercial activities. For instance, in India, groundwater is a vital water source across all sectors. However, groundwater availability faces significant risks due

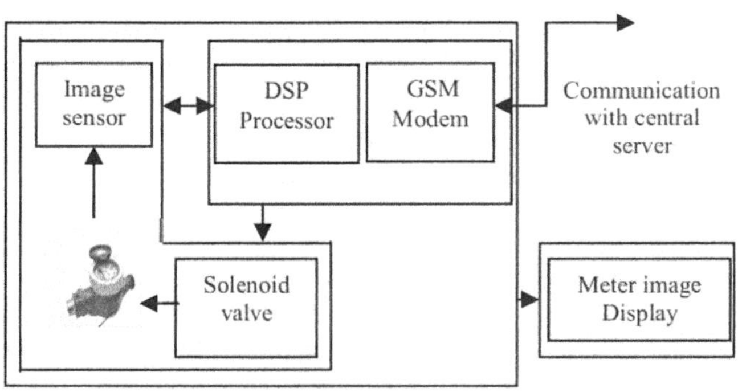

FIGURE 6.2 Intelligent water distribution system in which an image sensor captures the water-meter reading, a digital signal processor (DSP) interprets the image and controls a solenoid valve, and a GSM modem transmits consumption data to a central server. The meter image is displayed locally while enabling remote monitoring and control of water supply (Kashid & Pardeshi, 2014).

to climate unpredictability and variations. Therefore, it is essential to implement an effective water resource management strategy that can predict fluctuations in groundwater levels. Such predictions are necessary to ensure the sustainable utilisation of groundwater in both urban and rural areas. In the context of West Orissa, the water table experiences substantial periodic variations due to the demand for household water, irrigation, and commercial purposes. Pradhan et al. (2019) studied groundwater behaviour in the lower region of the Ganga-Ramganga inter-basin. They employed the co-active neuro-fuzzy inference system and radial basis function network (RBFN) to predict water table levels accurately. Aizebeokhai et al. (2016) utilised geo-electrical resistivity soundings, 2D geo-electrical resistivity, and time-domain IP imaging to characterise subsurface water and determine the aquifer level for assessing groundwater potential. Wang et al. (2006) developed a back-propagation neural network (BPNN) to stimulate the three-dimensional dispersion of NO_3-N concentrations in groundwater. They incorporated land cultivation data and site-specific hydrogeological information from Huantai County, China. In their work, Lee et al. (2018) employed artificial neural network (ANN) models and support vector machines to analyse groundwater productivity potential (GPP) in Boryeong City, Korea. Li et al. (2017), on the other hand, used the radial basis function neural network and Grey Model GM (1, 1) model to predict variations in groundwater levels in Longyan City, Fujian Province, South China. Al-Mahallawi et al. (2012) applied various neural network models, including multilayer perceptrons, RBF, generalised regression neural network, and a novel model called linear networks, to estimate groundwater contamination by nitrate.

Arora et al. (2023) conducted a comprehensive survey on RBFN, focusing on its development and applications. Agarwal et al. (2018) employed a BPNN as a classifier to detect faults and identify faulty poles. To efficiently control network congestion and stabilise queue length, a new active queue management controller was developed (Bisoy & Pattnaik, 2018). Ghorbani et al. (2017) generated a groundwater potential map using the weights-of-evidence and evidential belief function models based on a geographic information system (GIS) in Azna Plain, Iran. Li et al. (2017) extensively studied the latest advancements in RBFN association methods, focusing on fundamental concepts, mathematical algorithms, and engineering applications. Lee et al. (2012) utilised an ANN and GIS to map local groundwater productivity potential (GPP) in an area near the city of Pohang, Korea. Shamsuddin et al. (2020) illustrated the growth and utilisation of ANNs for predicting groundwater levels in two vertical boreholes in a confined aquifer near the Langat River. Affandi and Watanabe (2007) predicted the variation in daily groundwater levels using ANFIS and data mining models such as Levenberg-Marquardt and RBF. They aimed to monitor and analyse the changing patterns of groundwater levels. Ghose and Samantaray (2019) developed regression and BPNN models to predict sediment concentration. They explored the application of sensor networks and integrated neural network algorithms to monitor groundwater sustainability.

In agriculture, the irrigation system is a fundamental process that directly affects crop production by supplying water to the cultivated land. Traditionally, farmers had to physically visit their fields to assess the water requirements, which was time-consuming and labour-intensive, especially when dealing with multiple agricultural

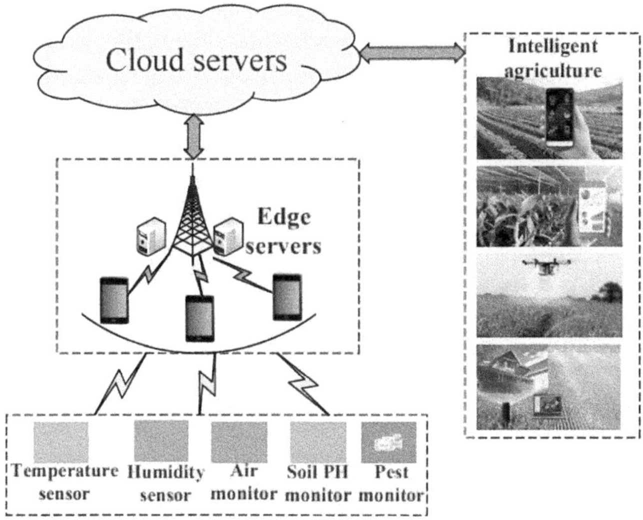

FIGURE 6.3 Smart water supply in agriculture, illustrating an IoT-based architecture in which field-deployed sensors (temperature, humidity, air quality, soil pH, and pest monitors) transmit data to edge servers and cloud platforms, where analytics support intelligent agriculture applications, such as precision irrigation, crop health monitoring, and remote farm management (Tong et al., 2021).

fields across different geographical areas. However, nowadays, farmers often juggle their agricultural activities with other occupations, making it challenging to carry out manual irrigation. Automating water irrigation systems provides a solution to make farmers' work easier. Sensor-based automated irrigation systems offer a promising approach where the presence of the farmer in the field is not necessary (Figure 6.3). Nowadays, the Internet is widely used, and farmers can monitor the irrigation status of their fields remotely. Mobile applications can fulfil this purpose by providing farmers with information about farm field moisture and temperature conditions. Farmers can use this data to decide whether to irrigate their fields, even from remote locations.

Smart farming, a concept gaining popularity in the agricultural industry, offers precise crop control, data collection, and automated farming techniques. Networked farms have numerous advantages to offer in terms of efficiency and productivity. Wakhare et al. (2020) developed an intelligent irrigation system controller that measures soil moisture and atmospheric temperature. The controller can activate or deactivate the water supply system based on these measurements. The soil moisture sensor detects the moisture level, indicating whether the soil is dry or wet. The moisture and temperature sensors are interfaced with a Peripheral Interface Controller (PIC) microcontroller. Based on the sensor values, the microcontroller activates a DC motor or pump by receiving instructions from a smartphone, which operates the pump via a relay. Water is a valuable resource that needs to be used efficiently, especially in agriculture, which consumes a significant amount of water. The IoT refers to the network of physical devices, vehicles, home appliances, and other items embedded with electronics, software, sensors, actuators, and network connectivity that enables these objects to connect and exchange data (see previous chapters). It allows remote sensing and control of objects across existing network infrastructure, facilitating the integration of the physical world into computer-based systems.

This integration results in improved efficiency, accuracy, economic benefits, and reduced human intervention in various domains, including agriculture (Wakhare et al., 2020).

6.2 COMPUTATIONAL METHODS

Computational methods use algorithms, mathematical models, and computer simulations to solve complex problems and analyse data in various scientific and engineering domains. These methods enable researchers and practitioners to handle large amounts of data, perform intricate calculations, and gain insights into complex systems. Computational methods encompass a wide range of techniques that leverage the power of computers to process and analyse data. Computational methods are continuously evolving, driven by advancements in computer hardware, algorithms, and data analysis techniques. They enable researchers to tackle complex problems, analyse big data sets, and gain insights that are otherwise challenging to obtain through traditional analytical approaches. Researchers and practitioners can explore intricate systems, optimise processes, and make informed decisions in various disciplines by leveraging computational methods. Computational methods are used in real-time monitoring and control systems for water supply networks. These methods enable the processing and analysis of real-time data, facilitating timely decision-making and control actions. For example, computational methods can identify abnormal system behaviour, detect leaks, predict equipment failures, and adjust control settings for pumps and valves. By utilising computational methods, intelligent computing and automation systems enable efficient and reliable real-time operation of water supply networks.

Computational methods provide powerful tools for solving complex problems and analysing data in various scientific and engineering domains. They are vital in implementing intelligent computing and automation techniques in water supply network management. These methods enable the analysis of large data sets, development of predictive models, optimisation of system performance, and decision support. Computational methods, such as optimisation algorithms, are employed to optimise the operation of water supply networks. These methods consider multiple variables, constraints, and objectives to find the most efficient and cost-effective solutions. For example, optimisation techniques can determine the optimal scheduling of pumps and valves, minimise energy consumption, optimise water allocation, and reduce water losses. By applying computational methods, intelligent computing systems optimise network operations in real time, ensuring water's reliable and efficient distribution (Gupta & Joshi, 2018). Computational methods, such as machine learning algorithms, are utilised in water supply network management to analyse vast amounts of data collected from sensors, meters, and other monitoring devices. These methods can identify patterns, correlations, and anomalies in the data, enabling the development of predictive models for various network parameters, such as demand forecasting, leak detection, and water quality prediction. By leveraging computational methods, intelligent computing systems can make accurate predictions and support proactive decision-making in water supply network management (Maier et al., 2014). Computational methods form the backbone of intelligent computing and automation in water supply network management. These methods can be broadly classified into two categories: numerical methods and simulation methods. These methods enhance our understanding of complex phenomena, support decision-making processes, and drive innovation in diverse fields.

6.2.1 Numerical Methods

Numerical methods involve solving mathematical equations and performing calculations using numerical approximations. These methods are instrumental when analytical solutions are not feasible or when dealing with complex systems. Numerical methods include finite element analysis, finite difference methods, optimisation algorithms, and numerical integration (Chapra & Canale, 2014). Numerical methods are extensively employed in physics, engineering, economics, and computational biology. They enable researchers to solve problems involving differential equations, optimisation, data fitting, and statistical analysis. Numerical methods provide approximate solutions with acceptable accuracy by discretising the problem domain and applying iterative algorithms.

6.2.2 Simulation Methods

Simulation methods involve creating computer models that mimic real-world systems or phenomena. These models incorporate mathematical and statistical representations of the underlying processes and are used to simulate and observe the system's behaviour under various conditions. Methods include Monte Carlo simulations, agent-based modelling, and system dynamics modelling. Simulation methods are widely employed in computer science, operations research, economics, and social sciences. They allow researchers to study complex systems, evaluate different scenarios, and make informed decisions based on the simulated outcomes. Simulation methods are instrumental when real-world experiments are expensive, time-consuming, or impractical.

6.2.2.1 Big Data Analysis

Big data analysis refers to examining large and complex data sets to extract valuable insights, patterns, and trends, and it involves employing computational methods and advanced analytics techniques to handle massive volumes of structured and unstructured data. Big data analysis has become increasingly important in various fields, including business, science, healthcare, and government, as it enables organisations to make data-driven decisions and gain a competitive edge (Chen et al., 2014). Big data analysis involves collecting, storing, processing, and analysing large and diverse data sets that exceed the capabilities of traditional data processing systems (Davenport & Dyche, 2013).

These data sets, often referred to as big data, are characterised by the three Vs:

- Volume (large amounts of data),
- Velocity (high speed of data generation), and
- Variety (different types and formats of data).

Big data analysis aims to derive meaningful insights, uncover hidden patterns, and make accurate predictions by applying computational methods, statistical techniques, machine learning algorithms, and data visualisation tools to these massive data sets (Manyika et al., 2011).

6.2.2.2 Challenges in Big Data Analysis

Big data analysis poses several challenges due to the volume, velocity, and variety of data. Some key challenges include data storage and management, data quality and veracity, data privacy and security, scalability of computational infrastructure, and computational efficiency of analysis algorithms. Addressing these challenges requires the development of robust data management systems, scalable computing frameworks, and advanced analytics techniques capable of handling large-scale data sets efficiently.

6.2.2.3 Applications of Big Data Analysis

Big data analysis finds applications in various domains. In business, it is used for customer analytics, market segmentation, fraud detection, and supply chain optimisation. In healthcare, big data analysis enables personalised medicine, disease surveillance, and drug discovery. In scientific research, it supports genomics, climate modelling, and particle physics. Additionally, big data analysis is applied in social media analysis, sentiment analysis, smart cities, and cybersecurity, among other areas. The insights derived from big data analysis help organisations improve decision-making, enhance operational efficiency, and drive innovation.

In summary, big data analysis involves the extraction of valuable insights from large and complex data sets through computational methods and advanced analytics techniques. It is a multidisciplinary field that addresses the challenges of managing and analysing massive volumes of data. Big data analysis has wide-ranging applications and empowers organisations to make informed decisions, gain a competitive advantage, and drive innovation in various industries.

6.3 WATER SUPPLY NETWORK DATA MANAGEMENT

Water supply network data management refers to the processes and systems involved in collecting, storing, organising, and utilising data related to water supply networks. It encompasses handling various types of data, including hydraulic, operational, customer, and sensor data, to ensure efficient management and operation of water supply systems. Water supply data collection is crucial in water supply network data management. It involves acquiring data from various sources, such as flow meters, pressure sensors, water quality sensors, and customer billing systems. These data sources provide information about water demand, network performance, asset conditions, and customer usage patterns. Data collection methods may include manual data entry, automated telemetry systems, and remote sensing technologies. Timely and accurate data collection ensures the availability of reliable data for analysis and decision-making (Eksioglu et al., 2018).

6.3.1 Water Supply Network Data Storage

Efficient data storage is essential for water supply network data management. Large volumes of data need to be securely stored and organised to facilitate easy access and retrieval. Database management systems are commonly used to store water supply network data. These systems provide structured storage, efficient querying capabilities, and data integrity. Additionally, advancements in cloud computing offer scalable and cost-effective storage solutions for water supply network data (Wu et al., 2016).

Cheng and Xiaobo (2009) performed Automatic Control Research on the Water Curtain System of an Underground Water Seal Oil Cavern. They concluded that a database is the central storage and processing unit for data information. It performs various functions, such as data transfer, updating, interpolation fitting, and query execution. In optimising the effectiveness of a water sealing system, the water curtain system's water pressure can be continuously adjusted based on a set of input parameters. Data related to the water curtain system, including the pore pressure of the water curtain holders, water injection pressure goodwell points, oil reservoir level conditions, and corresponding water inflow, are transferred into a database system. During the operational phase of oil reservoirs, the oil storage conditions play a significant role, with oil levels being the primary factor affecting the settings of the water curtain system. Therefore, the oil pressure of the cavern can be designated as the primary key.

Dynamic data tables can be established based on imported values. These dynamic data tables operate as follows: new feasible data is constantly added to the table, and the mark associated with identical data is incremented. Over time, data that has not been used for an extended period will be removed from the table after its mark decreases to a negative value. Data with higher scores will be preferred for future use. Additionally, dynamic data tables involve the creation of subsidiary tables alongside the primary table. The data from the primary table can be learned and trained using a neural network model, and the new data obtained from this process is inserted into the subsidiary table. A data query is performed when an input parameter enters the control system. Suppose the requested data does not exist in the primary table. In that case, it can be linked to the subsidiary table to search for related data, and this data is then inserted into the primary table with a mark of zero. If the data is proven valid, its mark is increased by one. Conversely, all data with a zero mark is removed from the primary table.

Water supply network data management involves integrating data from various sources and formats to create a unified view of the system. The integration allows the correlation of data from different sensors, meters, and databases to understand the network's behaviour comprehensively. Data integration techniques may include data transformation, normalisation, and aggregation. The integration enables data-driven analysis, modelling, and decision support for water supply network management (Liu et al., 2014).

6.3.2 Water Supply Network Data Compressions

Water supply network data compression refers to reducing the size of data related to water supply networks while preserving essential information and minimising the loss of data quality. Data compression techniques are applied to minimise the storage requirements, transmission bandwidth, and processing time associated with water supply network data. Data compression encodes data in a more compact representation to reduce the storage space or bandwidth needed for transmission. In the context of water supply networks, data compression techniques aim to reduce the size of various types of data, including sensor readings, telemetry data, and network parameters, without significantly compromising the information content (Sayood, 2017).

6.3.2.1 Techniques of Water Supply Network Data Compressions

Compression methods can be categorised into lossless and lossy compression techniques (Salomon, 2004; Sayood, 2017).

6.3.2.1.1 Lossless Compression

Fewer compression techniques ensure that the compressed data can be fully reconstructed to its original form without losing information. Lossless compression is suitable for water supply network data that requires precise preservation, such as hydraulic measurements or control system data (Salomon, 2004). Popular lossless compression algorithms used in water supply network data management include:

- Run-Length Encoding (RLE): RLE replaces repetitive data sequences with a single data value and a count of repetitions.
- Huffman Coding: Huffman coding assigns shorter codes to more frequently occurring data patterns, resulting in efficient representation.
- Arithmetic Coding: Arithmetic coding converts data into fractional numbers within a specific range, achieving higher compression ratios.

6.3.2.1.2 Lossy Compression

Lossy compression techniques selectively discard specific, less significant details in the data to achieve higher compression ratios. Lossy compression is often applied to water supply network data where the minor loss of precision is tolerable, such as sensor data for trend analysis or graphical representations. Standard lossy compression methods used in water supply network data management include:

- Transform-Based Compression: Techniques like Discrete Cosine Transform and Wavelet Transform are used to convert data into frequency or scale domains, allowing the removal of less perceptually important components.
- Quantisation: Quantisation reduces the precision of data by mapping values to a smaller set of representable values, resulting in a loss of information.
- Prediction-Based Compression: Prediction algorithms exploit the correlation between adjacent data points to estimate and encode the differences, reducing the amount of data stored or transmitted.

It is important to note that the selection of compression techniques depends on the specific requirements of the water supply network data, the trade-off between compression ratios and data quality, and the computational resources available for compression and decompression processes.

6.3.3 Water Supply Network Data Aggregation

Water supply network data aggregation combines multiple data points or measurements into a single representative value or summary statistic. It involves grouping and condensing data to reduce volume while capturing essential information. Data aggregation is helpful in water supply network management to simplify data analysis, reduce storage requirements, and improve computational efficiency. Data aggregation in water supply network management involves combining individual data points or measurements from various sources, such as sensors, meters, or customer records, into aggregated values or statistics. Aggregation can be performed over different temporal or spatial scales, depending on the specific requirements of the analysis or decision-making process. Data aggregation aims to summarise and condense data while retaining important trends, patterns, or characteristics (Gupta & Joshi, 2018).

6.3.3.1 Temporal Data Aggregation

Temporal data aggregation involves consolidating data over specific time intervals, such as minutes, hours, days, or months (Maier et al., 2014). Aggregating data over time can help reduce data volume while retaining important temporal trends and patterns. Common techniques for temporal data aggregation include:

- Averaging: Calculating the average value of data points within a time interval.
- Summarisation: Generating summary statistics, such as minimum, maximum, or median values, within a time interval.
- Sampling: Selecting a subset of data points at regular intervals to represent the overall trend.

6.3.3.2 Spatial Data Aggregation

Aggregation involves combining data from multiple locations or spatial units into a representative value or statistics (Kumar et al., 2019). It is particularly useful when dealing with large water supply networks with numerous monitoring points. Common techniques for spatial data aggregation include:

- Zonal Aggregation: Grouping data based on predefined spatial zones or districts and calculating aggregated values for each zone.
- Grid-Based Aggregation: Dividing the network into a grid and summarising data within each grid cell.
- Interpolation: Estimating values for unmonitored locations based on neighbouring observations and aggregating the interpolated values.

Data aggregation can help reduce the volume of data, simplify data analysis, and improve computational efficiency in water supply network management. However, it is crucial to consider the trade-off between data granularity and the level of detail required for specific analyses or decision-making processes.

6.4 CHALLENGES AND PROGRESS

6.4.1 Challenges of Intelligent Computing

Deep learning, a specific example of AI, has faced diverse challenges, especially in autonomy, evolvability, interpretability, and generality. AI technology, when compared to human intelligence, always falls short but only thrives in some specific cases and tasks. There is a lot to do to achieve successful and universal AI. Moreover, a transition from data-driven intelligence to various other forms of intelligence like autonomous intelligence, cognitive intelligence, human-machine fusion intelligence, and perpetual intelligence invokes unavoidable hurdles in its theoretical and technical background. Digitalisation has brought about an unprecedented surge in applications, connections, data generation, terminals, and users, requesting a computational capacity. For example, the computational power required for any AI doubles in 100-day intervals, projected to increase above a millionfold in five years (Zhu et al., 2023). An inevitable challenge is keeping up with such growth in computational capacity, particularly with the slowing down of Moore's law. In addition, in an intelligent society, this complex task necessitates the functional integration of various computing resources. There is often no alignment between traditional hardware models and intelligent algorithms, limiting software development.

Intelligent computing has the potential to revolutionise water supply network management by enabling efficient operation, proactive maintenance, and improved decision-making. However, their implementation faces several challenges. Some of the key challenges of Intelligent Computing Automation include (Wang et al., 2015; Hassanzadeh et al., 2020):

1. Data Management and Integration: Challenge: Water supply networks generate massive amounts of data from various sources, including sensors, meters, and SCADA systems. Managing and integrating this heterogeneous data in real-time can be challenging due to data quality, data compatibility, and data privacy.
2. System Complexity and Dynamics: Challenge: Water supply networks are complex systems with dynamic behaviour influenced by demand fluctuations, pipe network topology, and hydraulic interactions. Developing intelligent computing and automation systems that can effectively model and control these complex dynamics is a challenge.
3. Cybersecurity and Resilience: Challenge: With the increasing integration of intelligent computing and automation technologies, the vulnerability of water supply networks to cyber-attacks and system failures has become a major concern. Ensuring the cybersecurity and resilience of these systems is a significant challenge.

4. Human-Technology Interaction: Challenge: The successful implementation of intelligent computing and automation technologies relies on effective human-technology interaction. Key challenges are overcoming resistance to change, ensuring user acceptance, and providing adequate training and support for operators and staff.
5. Cost and Return on Investment: Challenge: The cost of implementing intelligent computing and automation technologies can be a barrier to their widespread adoption. Water utilities may face budget constraints and challenges in demonstrating the return on investment (ROI) for implementing these technologies.

6.4.2 Progress Made in Addressing the Challenges

Significant progress has been made in addressing the challenges associated with intelligent computing in water supply network management. Ongoing research, technological advancements, and collaboration between researchers, technology providers, and water utilities are driving the adoption and implementation of these technologies, leading to more efficient and sustainable water supply network management (Hassanzadeh et al., 2020; Wang et al., 2018; Zapata-Sierra et al. 2023).

1. Advances in data management techniques, such as big data analytics, cloud computing, and data integration platforms, have facilitated the handling and integration of diverse data sources. Data quality assurance methods, data cleansing algorithms, and standardised data formats have been developed to improve the accuracy and reliability of data used for intelligent computing and automation in water supply network management.
2. Researchers have made progress in developing advanced modelling and control algorithms that can handle the complexity and dynamics of water supply networks. Machine learning techniques, such as neural networks and genetic algorithms, have been applied to model and predict system behaviour. Advanced control strategies, such as model predictive control, have optimised system performance and addressed dynamic operational challenges.
3. Efforts have been made to enhance the cybersecurity and resilience of intelligent computing and automation systems. This includes the implementation of secure communication protocols, encryption techniques, intrusion detection systems, and redundant architectures. Cybersecurity training and awareness programmes have also been conducted to educate water utility staff about potential threats and best practices for maintaining a secure and resilient system.
4. Progress has been made in addressing the human-technology interaction challenges. Human-centred design principles have been employed to develop user-friendly interfaces and intuitive DSS. Training programmes, workshops, and collaboration between technology providers and water utilities have helped operators and staff understand and utilise intelligent computing and automation technologies.

5. The decreasing cost of computing hardware, data storage, and communication technologies has made intelligent computing and automation more affordable. Additionally, studies and case examples demonstrating the economic benefits, such as energy savings, reduced water losses, and improved operational efficiency, have helped utilities justify the investment and realise the long-term ROI.

6.5 CHAPTER SUMMARY AND CONCLUSION

Intelligent computing has helped advance AI significantly by fusing AI with computer technologies. Its end goal is to design computational activities that make the most use of available resources while minimising waste and maximising efficiency. Maintaining a reliable and sustainable water distribution system relies heavily on well-managed water supply networks. Computing power, data analytics, and control system advancements have all contributed to a sea of change in the industry. Predictive analytics in water supply network management is made possible using intelligent computing methods like machine learning and AI. These algorithms analyse previous data to help predict future needs, anticipate probable breakdowns, and optimise maintenance schedules to uncover trends, correlations, and anomalies. The potential for water waste and infrastructure damage may be mitigated with the use of fault detection algorithms, which can automatically identify system anomalies. By facilitating effective operation, preventative maintenance, and enhanced decision-making, intelligent computing has the potential to transform the administration of water supply networks completely. To manage the complexity and dynamism of water supply networks, researchers are developing data management strategies, cutting-edge modelling and control algorithms, cybersecurity measures, and methods for enhancing human-technology interaction.

REFERENCES

Affandi, A. K., & Watanabe, K. (2007). Daily groundwater level fluctuation forecasting using soft computing technique. *Natural Science*, *5*(2), 1–10.

Agarwal, S., Swetapadma, A., & Panigrahi, C. (2018). An improved method using artificial neural network for fault detection and fault pole identification in voltage source converter-based high-voltage direct current transmission lines. *Arabian Journal for Scie nce and Engineering*, *43*(8), 4005–4018.

Aizebeokhai, A. P., Oyeyemi, K. D., & Joel, E. S. (2016). Groundwater potential assessment in a sedimentary terrain, southwestern. *Nigeria Arabian Journal of Geosciences*, *9*(7), 496.

Al-Mahallawi, K., Mania, J., Hani, A., & Shahrour, I. (2012). Using of neural networks for the prediction of nitrate groundwater contamination in rural and agricultural areas. *Environmental Earth Sciences*, *65*(3), 917–928.

Arora, S., Ayebire, A.-M., & Manshahia, M. S. (2023). A survey on optimization of multi-criteria RBFN. In P. Vasant, M. Shamsul Arefin, V. Panchenko, J. J. Thomas, E. Munapo, G. W. Weber & R. Rodriguez Aguilar (Eds.), *Intelligent Computing and Optimization* (Lecture Notes in Networks and Systems (855: 205–213). Springer Nature Switzerland AG. https://doi.org/10.1007/978-3-031-50151-7_20

Bisoy, S. K., & Pattnaik, P. K. (2018). An AQM controller based on feed-forward neural networks for stable internet. *Arabian Journal for Science and Engineering*, *43*(8), 3993–4004.

Chapra, S. C., & Canale, R. P. (2014). *Numerical Methods for Engineers* (7th ed.). McGraw-Hill.

Chen, M., Mao, S., & Liu, Y. (2014). Big data: A survey. *Mobile Networks and Applications*, *19*(2), 171–209.

Cheng, L., & Xiaobo, X. (2009). Automatic control research on water curtain system of underground water seal oil cavern. In *2009 Second International Conference on Intelligent Computation Technology and Automation* (Vol. 1, pp. 758–761). IEEE.

Davenport, T. H., & Dyche, J. (2013). Big data in big companies. *International Institute for Analytics*, *3*, 1–31.

Eksioglu, S. D., Kisi, O., & Ozturk, O. (2018). A review of data-driven modelling approaches for water supply and distribution systems. *Journal of Hydrology*, *563*, 1096–1111.

Ghorbani Nejad, S., Falah, F., Daneshfar, M., Haghizadeh, A., & Rahmati, O. (2017). Delineation of groundwater potential zones using remote sensing and GIS-based data-driven models. *Geocarto International*, *32*(2), 167–187.

Ghose, D. K., & Samantaray, S. (2019). Integrated sensor networking for estimating groundwater potential in scanty rainfall region: Challenges and evaluation. *Computational Intelligence in Sensor Networks*, *776*, 335–352.

Gupta, A., & Joshi, D. (2018). Intelligent computing for water distribution network: A review. *International Journal of Engineering and Technology*, *7*(4.30), 420–424.

Hassanzadeh, A., Rasekh, A., Galelli, S., Aghashahi, M., Taormina, R., Ostfeld, A., & Banks, M. K. (2020). A review of cybersecurity incidents in the water sector. *Journal of Environmental Engineering*, *146*(5), 03120003.

Kashid, S. G., & Pardeshi, S. A. (2014). A survey of water distribution system and new approach to intelligent water distribution system. In *First International Conference on Networks & Soft Computing IEEE* (pp. 339–344).

Kumar, A., Pant, M., & Babu, P. V. S. (2019). A comprehensive review of data-driven models for water distribution system analysis. *Water Resources Management*, *33*(6), 2131–2151.

Lee, S., Hong, S. M., & Jung, H. S. (2018). GIS-based groundwater potential mapping using artificial neural network and support vector machine models: The case of Boryeong city in Korea. *Geocarto International*, *33*(8), 847–861.

Lee, S., Song, K. Y., Kim, Y., & Park, I. (2012). Regional groundwater productivity potential mapping using a geographic information system (GIS) based artificial neural network model. *Hydrogeology Journal*, *20*(8), 1511–1527.

Li, Z., Yang, Q., Wang, L., & Martín, J. D. (2017). Application of RBFN network and GM (1, 1) for groundwater level simulation. *Applied Water Science*, *7*(6), 3345–3353.

Lian, Y., Ren, Y., & Wu, J. (2019). A survey on smart water grid: From the perspective of communications and data analytics. *IEEE Access*, *7*, 57375–57388.

Liu, W., Fu, G., Liu, X., Zeng, Y., & Zeng, W. (2021). A review of intelligent fault diagnosis methods for water supply pipelines. *IEEE Access*, *9*, 13685–13703.

Liu, Y., Maier, H. R., Guo, H., Dandy, G. C., & Jin, J. (2014). Data integration for improved modelling of water distribution systems. *Environmental Modelling & Software*, *52*, 265–276.

Maier, H. R., Kapelan, Z., Kasprzyk, J., Kollat, J., Matott, L. S., Cunha, M. C., Dandy, G. C., Gibbs, M. S., Keedwell, E., Marchi, A., Ostfeld, A., Savic, D., Solomatine, D. P., Vrugt, J. A., & Zecchin, A. C. (2014). Evolutionary algorithms and other metaheuristics in water resources: Current status, research challenges and future directions. *Environmental Modelling & Software*, *62*, 271–299.

Manyika, J., Chui, M., Brown, B., Bughin, J., Dobbs, R., Roxburgh, C., & Hung Byers, A. (2011). *Big Data: The Next Frontier for Innovation, Competition, and Productivity*. McKinsey Global Institute.

Pradhan, S., Kumar, S., Kumar, Y., & Sharma, H. C. (2019). Assessment of groundwater utilization status and prediction of water table depth using different heuristic models in an Indian interbasin. *Soft Computing*, *23*(20), 10261–10285.

Salomon, D. (2004). *Data Compression: The Complete Reference* (4th ed.). Springer.

Shamsuddin, M. K. N., Kusin, F. M., Sulaiman, W. N. A., Ramli, M. F., Baharuddin, M. F. T., & Adnan, M. S. (2017). Forecasting of Groundwater Level using Artificial Neural Network by incorporating river recharge and riverbank infiltration. In *MATEC Web of Conferences* (Vol. 103, p. 04007). EDP Sciences.

Sayood, K. (2017). *Introduction to Data Compression* (5th ed.). Morgan Kaufmann.

Tong, Z., Ye, F., Yan, M., Liu, H., & Basodi, S. (2021). A survey on algorithms for intelligent computing and smart city applications. *Big Data Mining and Analytics*, *4*(3), 155–172.

Valeyudhan, S., Franchini, M., & Savic, D. (2022). Water distribution systems analysis: State-of-the-art review. In *Water Distribution Systems Analysis Symposium*, Cincinnati, Ohio, USA. https://doi.org/10.1109/ACCESS.2022.3208142

Wakhare, P. B., Neduncheliyan, S., & Sonawane, G. S. (2020). Automatic irrigation system based on the Internet of Things for crop yield prediction. In *2020 International Conference on Emerging Smart Computing and Informatics (ESCI)* (pp. 129–132). IEEE.

Wang, M. X., Liu, G. D., Wu, W. L., Bao, Y. H., & Liu, W. N. (2006). Prediction of agriculture-derived groundwater nitrate distribution in North China Plain with GIS-based BPNN. *Environmental Geology*, *50*(5), 637–644.

Wang, Z., Song, H., Watkins, D. W., Ong, K. G., Xue, P., Yang, Q., & Shi, X. (2015). Cyber-physical systems for water sustainability: challenges and opportunities. *IEEE Communications Magazine*, *53*(5), 216–222.

Wu, J., Cui, H., & Wu, L. (2016). A cloud computing-based framework for water environment monitoring in smart cities. *Journal of Hydroinformatics*, *18*(6), 1048–1059.

Zapata-Sierra, A. J., Salmerón-Manzano, E., Alcayde, A., Zapata-Castillo, M. L., & Manzano-Agugliaro, F. (2023). The scientific landscape of smart water meters: A comprehensive review. *Water*, *16*(1), 113.

Zhu, S., Yu, T., Xu, T., Chen, H., Dustdar, S., Gigan, S., Gunduz, D., Hossain, E., Jin, Y., Lin, F., Liu, B., Wan, Z., Zhang, J., Zhao, Z., Zhu, W., Chen, Z., Durrani, T. S., Wang, H., Wu, J., Zhang, T., & Pan, Y. (2023). Intelligent computing: The latest advances, challenges, and future. *Intelligent Computing*, *3*, Article 0006. https://doi.org/10.34133/icomputing.0006

7 Automation in the Water Supply Network Management

7.1 INTRODUCTION

Automation is a collection of a whole set of theories and methods to make a system work automatically as intended, in our case, the urban water supply system. The monitoring system for urban water supply has garnered significant interest in recent years. Urban water supply networks serve as the crucial connection between the provision of drinking water and the end-users who consume it (Panchal et al., 2014). The water distribution supply systems play a critical role in ensuring the continuous provision of water, maintaining water quality control, and monitoring the technological process parameters. According to Olsson (2021), a crucial feature of automation in water supply network management is the feedback principle where a sensor measures a specific variable, for example, a concentration; a computer tests that the measurement is valid; a computer algorithm calculates and decides what should be corrected; and a pump or valve or some other device transforms the decision to action. Incorporating an automated system into the water distribution network ensures the timely updating of the renovated water supply infrastructure in metropolitan areas. This system provides novel methods for monitoring and optimising the utilisation of water resources and technological equipment (Gouthaman et al., 2011). The implementation of automation coupled with instrumentation and control is presently observed across several components of the urban water system, including water treatment facilities, water distribution networks, sewer networks, and wastewater treatment plants. Integrated urban water management is often regarded as an enabling technology. Even at the infancy stage across the globe, the expeditious advancement of sensors, instruments, and communication systems has facilitated the acquisition of data in real time from various process units throughout the entirety of the urban water system (Yuan et al., 2019).

7.2 REAL-TIME CONTROL

One of the key applications of automation in water supply network management is real-time monitoring and control. By deploying sensors and data acquisition systems, operators can gather comprehensive data on flow rates, pressure levels, water quality, and equipment status. This data is the foundation for decision-making, enabling operators to identify anomalies, predict system behaviour, and take timely actions. Automation systems, driven by advanced control algorithms, facilitate

DOI: 10.1201/9781003633839-7

the optimisation of pumps, valves, and other network components in response to real-time conditions, resulting in improved operational efficiency and reduced water loss and energy consumption.

7.3 ADVANCED CONTROL SYSTEM TECHNOLOGIES

Advanced control system technologies use sophisticated and advanced techniques to improve the control and operation of various systems, including industrial processes, power grids, transportation networks, and environmental systems. These technologies leverage computational power, data analytics, machine learning, and automation to enhance system performance, efficiency, and reliability. A significant example is the model predictive control (MPC) described by Maciejowski (2002) as a control strategy that utilises mathematical models of the system to predict its future behaviour and optimise control actions accordingly. MPC considers constraints, system dynamics, and objectives to generate optimal control actions over a defined time horizon. It has been widely applied in various industries, including chemical process control, power systems, and building automation, to improve control performance and handle complex system behaviours. Another technology of this kind is fuzzy logic control, which Ross (2010) explained as a control approach that handles imprecise or uncertain information using fuzzy sets and linguistic variables. Fuzzy logic control (FLC) allows for a more flexible and intuitive representation of control rules, enabling the control system to handle complex or nonlinear system dynamics. FLC has found applications in various fields, including automotive systems, consumer electronics, and intelligent transportation systems. Distributed control systems (DCS) are advanced control architectures consisting of a network of distributed controllers, sensors, and actuators. A DCS allows for decentralised control, enabling local decision-making and coordination among multiple control units (Luyben, 2013). DCS are extensively used in large-scale industrial processes, power plants, and smart grids to improve system reliability, scalability, and fault tolerance. Intelligent control integrates artificial intelligence (AI) techniques into control systems such as machine learning, expert systems, and neural networks. It allows the control system to learn, adapt, and optimise control actions based on historical data and real-time inputs. Intelligent control has applications in various domains, including robotics, autonomous vehicles, and smart buildings, enabling systems to make data-driven decisions and improve performance (Korbicz et al., 2011). These advanced control system technologies offer significant benefits in system performance, efficiency, and reliability. They enable more accurate and adaptive control, leading to improved productivity, energy savings, and operational cost reduction in various applications.

7.3.1 Adaptive Control Technologies

Adaptive control technologies refer to control techniques that adjust control parameters in real time based on system changes and uncertainties. These technologies enable control systems to adapt and optimise their performance to changing conditions, variations, or disturbances. Adaptive control plays a crucial role in applications where system dynamics are uncertain, varying, or challenging to model accurately.

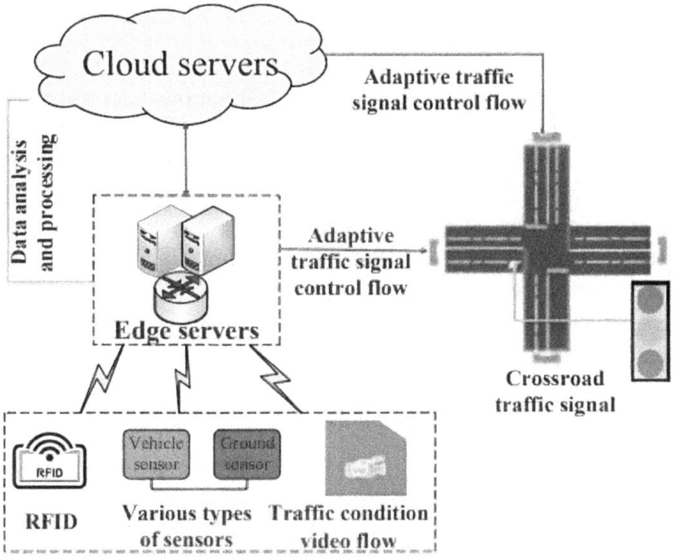

FIGURE 7.1 Workflow of an Adaptive Traffic Signal Control System illustrating real-time data collection, traffic pattern analysis, and dynamic signal optimisation to improve intersection efficiency and reduce congestion (adapted from Tong et al., 2021).

Adaptive control techniques, such as adaptive Proportional–Integral–Derivative (PID) control, model reference adaptive control (MRAC), and self-tuning control, have been applied in diverse applications, including aerospace, robotics, and manufacturing, to enhance system performance in the presence of uncertainties. An example is in its application to the traffic control signal, as depicted in Figure 7.1.

MRAC is a popular adaptive control approach that compares the system's actual output with a desired reference model's output and adjusts control parameters to minimise the error. MRAC uses a mathematical model of the system to estimate its dynamics and adaptively updates the control law to track the desired performance. MRAC has been widely applied in aerospace, robotics, and industrial processes where system dynamics may change due to variations or uncertainties (Ioannou & Sun, 2012). On the other hand, self-tuning control (STC) is an adaptive control technique that continuously estimates the system's dynamics and adjusts control parameters based on observed system behaviour. STC algorithms typically employ identification methods, such as recursive least squares or instrumental variable methods, to update the model parameters and adapt the control law. STC has been successfully applied in various industrial processes, including chemical plants and power systems, to handle uncertainties and improve control performance (Astrom & Wittenmark, 2013). Moreover, reinforcement learning (RL) is an adaptive control technique that utilises machine learning algorithms to learn control policies through trial and error. RL agents interact with the environment, receive feedback through rewards or penalties, and update their control strategies to maximise cumulative rewards. RL has gained attention in applications, such as robotics, autonomous vehicles, and game playing, where the system dynamics and optimal control strategies may be unknown or complex (Sutton & Barto, 2018).

Furthermore, neural network control utilises artificial neural networks to approximate complex system dynamics and adapt control actions accordingly. Neural networks are trained to learn the mapping between system inputs and desired control outputs, enabling adaptive control even in the presence of uncertainties or nonlinearities. Neural network control has applications in various domains, including process control, robotics, and intelligent transportation systems (Korbicz et al., 2011). Adaptive control technologies offer significant advantages in scenarios where system dynamics are uncertain, changing, or challenging to model accurately. They provide the ability to adapt control strategies in real time, improving control systems' performance, robustness, and stability. By continuously monitoring and adjusting control parameters, adaptive control technologies enable systems to handle disturbances, variations, and uncertainties, enhancing operational efficiency and reliability.

7.3.2 Model Predictive Control

MPC is an advanced control strategy that utilises a mathematical model of the system to predict its future behaviour and optimise control actions over a finite time horizon. MPC is widely used in various industries and applications where control performance, constraints, and dynamic behaviour are critical factors. A MPC consists of the following key components:

- System Model: A mathematical model that represents the dynamics of the system being controlled. This model can be based on first principles or derived from system identification techniques.
- Objective Function: A performance metric or cost function quantifying the desired control objectives. The objective function typically includes terms related to setpoint tracking, control effort, and constraint violations.
- Prediction Horizon: The length of the interval over which predictions are made. The prediction horizon determines how far into the future the control actions are optimised.
- Control Horizon: The time interval over which control actions are applied. The control horizon specifies the duration for which control inputs are implemented.
- Constraints: Limits or boundaries on system variables must be satisfied during control. Constraints can include physical limitations, safety limits, or operational constraints.
- Optimisation Algorithm: An algorithm that solves an optimisation problem to determine the optimal control actions over the prediction horizon. The optimisation considers the system model, objective function, constraints, and current system state.

7.3.2.1 Common Applications of Model Predictive Control

MPC finds applications in various industries and domains, ranging from its importance in building automation, process control, autonomous vehicles, and power systems, as shown in Figure 7.2. MPC is widely used in chemical processes, oil refineries, and manufacturing systems to regulate process variables, optimise operations, and handle constraints, such as equipment limits and product quality specifications. It is also applicable in power systems applied in power generation, transmission, and distribution systems to manage grid stability, load balancing, and renewable energy integration. It enables optimal power flow, voltage regulation, and grid frequency control. MPC is employed in building control systems for heating, ventilation, and air conditioning (HVAC) to optimise energy consumption, maintain comfort levels, and manage demand response. Also, MPC is utilised in autonomous vehicle control systems for trajectory planning, obstacle avoidance, and efficient energy management (Camacho & Bordons, 2007).

MPC offers optimisation control actions over a finite time horizon, considering future system behaviour, constraints, and control objectives. As revealed in Figure 7.3,

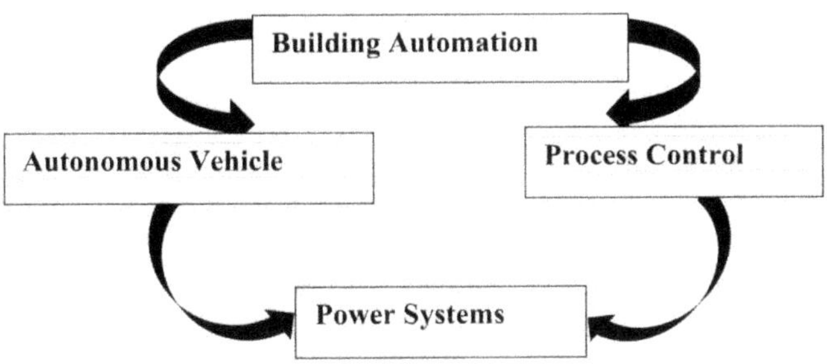

FIGURE 7.2 Model Prescriptive Control applications span multiple domains, including building automation for energy-efficient HVAC and lighting control, industrial process control for maintaining operational stability, autonomous vehicle navigation through real-time trajectory planning, and power systems for enhancing grid reliability and optimising energy distribution.

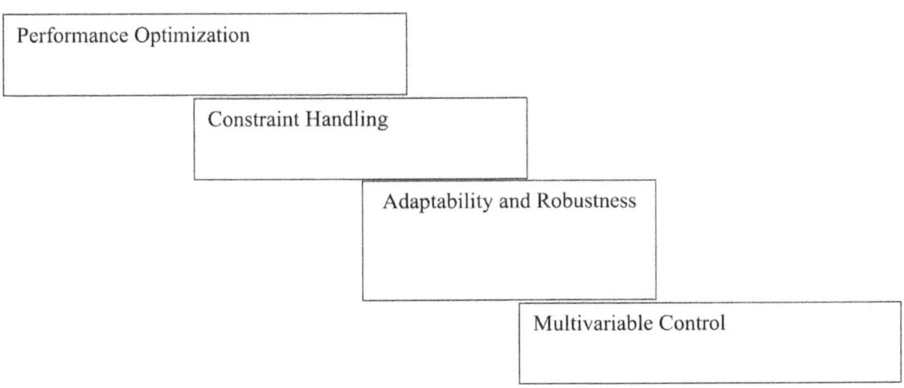

FIGURE 7.3 This figure illustrates the key benefits of model predictive control (MPC), including improved handling of system constraints, multivariable control, adaptability and robustness, enhanced stability, and superior optimisation performance compared to the conventional control method.

this enables improved tracking of setpoints, efficient utilisation of resources, and better control performance. It can handle constraints explicitly, ensuring that the control actions stay within the specified limits. It can handle hard constraints on system variables and ensure safe and stable operations. MPC can handle system uncertainties, disturbances, and time-varying dynamics by updating real-time control actions based on measured or estimated system states. This adaptability enhances the robustness and stability of the control system. It is also well-suited for multivariable control problems with interrelated inputs and outputs. It can optimise control actions across multiple variables and handle their interactions. In conclusion, MPC is an advanced control strategy that uses system models, optimisation algorithms, and predictions to determine optimal control actions over a finite time horizon. It finds applications in various industries where control of performance, constraints, and dynamic behaviour are crucial. MPC enables improved performance, constraint handling, adaptability, and multivariable control, leading to enhanced efficiency and operational effectiveness.

7.4 SENSOR INTEGRATION

Sensor integration refers to combining data from multiple sensors to obtain a comprehensive understanding of a system or environment. It involves the fusion and synchronisation of data from various sensors to enhance the accuracy, reliability, and richness of the information obtained. Sensor integration is crucial in numerous domains, including robotics, autonomous systems, environmental monitoring, healthcare, and smart cities. Sensor integration involves integrating data from multiple sensors, which can be of different types, modalities, or operating principles, to obtain a more complete and accurate representation of the monitored system or environment. Combining data from multiple sensors makes it possible to compensate for limitations or uncertainties associated with individual sensors, reduce noise, enhance spatial or temporal resolution, and extract more meaningful information for decision-making or control.

7.4.1 Benefits of Sensor Integration

Sensor integration offers several benefits in data acquisition, analysis, and decision-making processes. Some of these benefits, according to Julier and Uhlmann (2004), include:

- **Improved Accuracy and Reliability:** By combining data from multiple sensors, the accuracy and reliability of the obtained information can be enhanced. Sensor fusion can compensate for measurement errors, outliers, or limitations associated with individual sensors, leading to more reliable and robust data.
- **Enhanced Situational Awareness:** Sensor integration provides a more comprehensive view of the system or environment, enabling a better understanding of its state, dynamics, and behaviour. By fusing data from different sensors, a more complete and accurate system representation is possible, leading to improved situational awareness.

- **Redundancy and Fault Tolerance:** Sensor fusion allows for redundancy in data acquisition. In the event of sensor failures or malfunction, the integrated system can continue to operate by relying on the data from the remaining sensors. This improves the fault tolerance and resilience of the system.
- **Improved Decision-Making:** Sensor integration provides a richer and more nuanced understanding of the system, which can support better decision-making. By combining data from multiple sensors. It is possible to extract higher-level information, detect patterns or anomalies, and make informed decisions based on a more comprehensive data set.

7.4.2 Sensor Fusion Techniques

Various techniques are employed in sensor fusion to integrate data from multiple sensors. Some standard techniques, according to Liu and Xie (2011), include:

- Kalman Filtering: Kalman filtering is a recursive algorithm that estimates the state of a system based on noisy measurements. It combines predictions from a mathematical model of the system with sensor measurements to obtain an optimal estimate of the system's state.
- Particle Filtering: Particle filtering, also known as Monte Carlo filtering, is a non-parametric Bayesian filtering technique that uses a set of weighted samples (particles) to represent the system's state distribution. It provides a flexible framework for fusing data from multiple sensors.
- Dempster-Shafer Theory: Dempster-Shafer Theory is a mathematical framework for reasoning under uncertainty and combining evidence from multiple sources. It allows for the fusion of data from different sensors with varying levels of reliability or uncertainty.
- AI Techniques: Machine learning and AI techniques, such as neural networks and support vector machines, can fuse data from multiple sensors. These techniques teach the relationships and dependencies between sensor data and can effectively integrate information from different sources.

Sensor integration, in summary, plays a vital role in obtaining a comprehensive understanding of a system or environment by combining data from multiple sensors. It offers improved accuracy, enhanced situational awareness, redundancy, and better decision-making capabilities—various techniques, such as Kalman filtering. Dempster-Shafer theory, and AI are employed for sensor fusion, depending on the specific requirements and characteristics of the application domain. Sensor integration is a fundamental aspect of modern sensing systems and is crucial for enabling advanced functionalities in numerous fields.

7.5 APPLICATIONS OF ADVANCED CONTROL SYSTEMS TECHNOLOGY TO WATER SYSTEMS

Advanced control system technologies have numerous applications in water systems, enabling efficient and optimal operation, control, and management. The following includes applications of advanced control system technologies in water systems (Liu & Xie, 2011; Savic & Walters, 2015).

1. **Pumping Systems Optimisation:** Advanced control algorithms can optimise pumping systems in water distribution networks by adjusting pump speeds, pressure setpoints, and operational schedules based on real-time demand and network conditions. By implementing model-based predictive control or advanced optimisation techniques, energy consumption can be reduced, water pressure can be regulated, and system performance can be improved.
2. **Water Treatment Process Control:** Advanced control systems can enhance the efficiency and effectiveness of water treatment processes. By integrating real-time data from sensors and instruments, control algorithms can optimise chemical dosing, flow rates, and process parameters to achieve desired water quality standards. Advanced control strategies, such as MPC, can handle complex dynamics, uncertainties, and changing water quality conditions.
3. **Desalination Plant Control:** Desalination plants convert seawater into freshwater and require precise control to optimise energy consumption and production rates. Advanced control system technologies can be applied to desalination processes, such as reverse osmosis or multi-stage flash distillation. Control algorithms can regulate feedwater flow rates, pressure, and membrane cleaning cycles to improve efficiency and minimise energy consumption.
4. **Water Distribution System Control:** Advanced control systems are vital in managing water distribution networks. By utilising real-time sensor data, control algorithms can optimise valve operations, pressure control, and flow balancing to minimise water losses, maintain network stability, and ensure consistent water supply to consumers. Advanced control techniques, such as adaptive control or network-wide optimisation, can handle varying demand patterns and system uncertainties.
5. **Reservoir Operation and Water Resource Management:** Advanced control strategies can optimise reservoir operation and water allocation in multi-reservoir systems. Control algorithms can optimise reservoir releases, storage levels, and water allocation strategies by integrating data from weather forecasts, inflow predictions, and demand forecasts. This helps to meet water supply demands, manage drought conditions, and maintain ecological flow requirements.

These applications highlight the significance of advanced control system technologies in water systems, enabling efficient water management, energy savings, improved water quality, and enhanced system performance.

7.6 APPLICATION OF INTELLIGENT COMPUTING AND AUTOMATION TO HYDROELECTRICITY

The application of intelligent computing and automation in water supply network management extends beyond traditional water supply systems and can also be beneficial in hydroelectricity stations (Agarwal et al., 2019).

1. **Real-Time Monitoring and Control Intelligence:** Computing and automation technologies can monitor various parameters in hydroelectricity stations, such as water flow rates, turbine performance, and power generation. Real-time data from sensors and monitoring devices can be collected using intelligent algorithms to optimise turbines' operations, regulate water flow, and maximise power generation efficiency.
2. **Predictive Maintenance:** By leveraging intelligent computing techniques and machine learning algorithms, it is possible to analyse historical data from hydroelectricity stations to predict maintenance needs and prevent equipment failures. This proactive approach helps schedule maintenance activities, reduce downtime, and optimise the lifespan of critical components, such as turbines and generators.
3. **Fault Detection and Diagnosis:** Intelligent computing and automation can facilitate the detection and diagnosis of faults or abnormalities in the operation of hydroelectric systems. By analysing sensor data and using pattern recognition algorithms, deviations from normal operating conditions can be identified, allowing for timely interventions and preventing potential damage to equipment or disruptions in power generation.
4. **Optimisation of Power Generation:** Advanced control strategies, such as MPC, can optimise power generation in hydroelectricity stations. By considering variables such as water availability, turbine efficiency, and electricity demand, these techniques can dynamically adjust operational parameters to maximise power output while ensuring the stability and reliability of the power grid.
5. **Integration with Energy Management Systems:** Intelligent computing and automation technologies can be integrated with energy management systems to optimise hydroelectricity stations' overall performance and efficiency. By exchanging data and control signals with other renewable energy sources or energy storage systems, intelligent systems can facilitate load balancing, grid stability, and optimal utilisation of available resources.

Intelligent computing and automation technologies in hydroelectricity stations offer several advantages, including increased operational efficiency, improved maintenance practices, enhanced reliability, and optimised power generation. These technologies enable operators to make informed decisions, maximise system performance, and ensure the sustainable utilisation of water resources for electricity generation.

7.7 SENSOR INTEGRATION APPLICATION IN WATER SYSTEMS

Sensor integration has significant applications in water systems, enabling improved monitoring, management, and decision-making in various aspects of water resource management. Here are some specific applications of sensor integration in water systems (Maksimovic & Teixeira, 2017; Chen & Wang, 2018; Mohammadi, 2018; Skarlatos et al., 2021).

7.7.1 Water Quality Monitoring

Sensor integration allows for the simultaneous measurement of multiple water quality parameters, such as pH, dissolved oxygen, turbidity, conductivity, and chemical contaminants. Integrating data from different sensors enables a more comprehensive understanding of water quality, early detection of pollution events, identification of contamination sources, and timely response to ensure a safe and clean water supply.

7.7.2 Leak Detection and Pipe Monitoring

Integrating data from multiple sensors, such as acoustic sensors, pressure sensors, and flow meters, can facilitate the detection of leaks in water distribution networks. By analysing sensor data collectively, it is possible to identify abnormal patterns, localise leaks, and prioritise repair and maintenance activities. Sensor integration can also enable continuous monitoring of pipe conditions, detecting pipe failures or deterioration, and preventing costly water losses.

7.7.3 Flood Monitoring and Early Warning Systems

In flood-prone areas, sensor integration is crucial in monitoring water levels, river flows, and rainfall patterns. Integrating data from multiple sensors installed across river basins or flood-prone regions makes it possible to monitor and predict flood events in real time. This information can be used to issue early warnings, implement flood mitigation measures, and improve emergency response planning.

7.7.4 Water Resource Management

Sensor integration can support efficient water resource management by integrating data from various sources, such as weather stations, river gauges, soil moisture sensors, and water demand sensors. By combining this data, water managers can make informed decisions regarding water allocation, irrigation scheduling, reservoir operations, and drought management. Sensor integration enhances the accuracy of hydrological models and improves water availability predictions.

7.7.5 Water Energy Efficiency

Integrating sensors in water systems enables real-time monitoring of water consumption, pressure levels, and energy usage. By analysing sensor data, water utilities can identify areas of excessive water consumption, detect inefficiencies in pumping systems, optimise energy usage, and implement water conservation measures. Sensor integration helps achieve water-energy nexus goals by promoting sustainable water use practices.

These applications demonstrate the importance of sensor integration in enhancing water system monitoring, management, and sustainability. By integrating data from various sensors, water utilities and decision-makers can make more informed and efficient decisions, improving water quality, conservation, and resilience in water systems.

7.8 APPLICATION OF AUTOMATION TO WATER CORPORATIONS

Intelligent computing and automation technologies have been widely applied in water supply network management in water corporations in France. These technologies offer improved efficiency, enhanced decision-making, and optimised resource management. Some specific applications of intelligent computing and automation in water corporations in France:

1. **Smart Metering and Monitoring:** Intelligent computing and automation technologies enable the implementation of smart metering systems in water corporations. These systems utilise advanced sensors, communication networks, and data analytics to monitor water consumption patterns, detect leaks, and optimise water distribution. Real-time data from smart meters can be analysed to identify abnormal consumption patterns, enabling proactive leak detection and conservation measures.
2. **Predictive Maintenance:** By leveraging intelligent computing techniques, machine learning algorithms, and historical data, water corporations in France can predict maintenance needs and optimise their maintenance schedules. Predictive maintenance enables the proactive identification of potential equipment failures, reducing downtime and improving the overall reliability of water supply networks. This approach helps prioritise maintenance activities and allocate resources effectively.
3. **Demand Forecasting and Resource Optimisation:** Intelligent computing and automation technologies facilitate demand forecasting in water corporations. These technologies can accurately predict water demand by analysing historical consumption data, weather patterns, and other relevant factors. This information helps water corporations optimise resource allocation, such as adjusting pumping schedules, optimising storage capacity, and managing water supply in real time to meet demand while minimising wastage.
4. **Leakage Detection and Management:** Intelligent computing and automation technologies play a crucial role in detecting and managing leaks in water supply networks. These technologies utilise advanced data analytics, pressure sensors, and flow meters to identify leakage points accurately. By continuously monitoring water flow rates, pressure variations, and acoustic patterns, anomalies indicative of leaks can be detected promptly, enabling rapid response and minimising water losses.
5. **Decision Support Systems:** Intelligent computing and automation technologies enable the development of decision support systems for water corporations in France. These systems utilise algorithms and models to provide insights and recommendations for water management, such as optimising pumping schedules, predicting water quality issues, and assisting in infrastructure planning. Decision support systems help water corporations make informed decisions, improve operational efficiency, and enhance overall performance.

The application of intelligent computing and automation technologies in water corporations in France has proven to be effective in optimising water supply network management. These technologies enable water corporations to improve operational efficiency, enhance resource utilisation, reduce water losses, and provide better service to customers. By leveraging these technologies, water corporations can meet regulatory requirements, address sustainability goals, and ensure a reliable supply of clean water to communities.

7.8.1 Applications of Intelligent Computing and Automation to Smart Water Grid in Barcelona, Spain

Intelligent computing and automation technologies have been applied in various places for water supply network management, demonstrating their effectiveness in improving system efficiency, decision-making, and resource management. Intelligent computing and automation technologies were implemented in Barcelona's water supply network. The system incorporated smart meters, sensors, and advanced analytics to monitor water consumption, detect leaks, and optimise water distribution. Real-time data analysis enabled proactive leak detection, reduction of non-revenue water, and more efficient resource allocation. This Smart Water Grid implementation demonstrates the application of intelligent computing and automation technologies in water supply network management. The city of Barcelona implemented a comprehensive system to optimise water distribution, reduce water losses, and improve overall operational efficiency (Rodríguez-Molina et al., 2020). Some key aspects of the Smart Water Grid project in Barcelona include:

1. **Sensor Network:** A network of smart sensors was deployed throughout the water supply network to monitor various parameters in real time. These sensors collected water flow rates, pressure levels, and water quality data. The sensor network provided a continuous stream of data, enabling accurate and timely monitoring of the network's performance.
2. **Data Analytics:** Advanced data analytics techniques were utilised to analyse the collected data from the sensor network. Machine learning algorithms and predictive models were employed to detect anomalies, identify potential leaks, and predict demand patterns. The data analytics provided insights into the system's behaviour, facilitating informed decision-making and proactive maintenance strategies.
3. **Leak Detection and Management:** The Smart Water Grid project in Barcelona focused on reducing water losses through efficient leak detection and management. By analysing the data from the sensor network, the system could identify abnormal flow patterns and pinpoint potential leak locations. This enabled timely repairs and minimised water losses, leading to substantial improvements in water conservation.
4. **Pressure Management:** Intelligent computing and automation technologies were utilised to optimise pressure levels within the water supply network. By dynamically adjusting pressure settings based on demand patterns and network conditions, excessive pressure was reduced, minimising the risk of pipe bursts and leakage while conserving energy.

5. **Customer Engagement:** The Smart Water Grid project aimed to engage customers actively in water conservation efforts. Water consumption data was made available to consumers through web portals and mobile applications, enabling them to monitor their usage, set conservation targets, and receive personalised recommendations for efficient water use. This approach fostered a sense of ownership and encouraged consumers to participate actively in water conservation.

Implementing the Smart Water Grid in Barcelona resulted in significant improvements in water management, including reduced water loss, optimised resource utilisation, improved infrastructure maintenance, and increased customer awareness about water conservation. The project demonstrated the potential of intelligent computing and automation technologies in transforming water supply network management and promoting sustainability.

7.8.2 Applications of Intelligent Computing and Automation to Automated Pump Control in the Netherlands

According to Bakkar et al. (2013), an intelligent control system was applied to optimise the operation of pumps in a water supply network in the Netherlands. The system utilised data-driven models and real-time sensor data to adjust pump operation based on water demand, pressure levels, and energy efficiency. The automated control resulted in energy savings, improved pressure management, and reduced maintenance costs. Implementing automated pump control in the Netherlands is an important case study highlighting the application of intelligent computing and automation technologies in water supply network management. The project aimed to optimise pump operation, improve energy efficiency, and enhance the overall performance of the water supply system (Bakkar et al., 2013). Some of the key aspects of the automated pump control project in the Netherlands are discussed below:

1. **Real-Time Monitoring:** The water supply system was equipped with sensors and monitoring devices to collect real-time data on water levels, pressure, and flow rates. This data was continuously monitored to provide up-to-date information on system conditions and performance.
2. **Intelligent Control Algorithms:** This includes advanced control algorithms, such as MPC, optimised pump operation based on real-time data, and predefined performance objectives. These algorithms consider factors like water demand, pressure levels, and energy consumption to determine the optimal operation of pumps.
3. **Energy Efficiency:** The automated pump control system improves energy efficiency by adjusting pump speeds and schedules based on the current demand and system conditions. The project aimed to minimise energy consumption by optimising pump operation while maintaining adequate water supply and pressure levels.

4. **Remote Monitoring and Control:** The system allowed for remote monitoring and control of the pumps. Operators could access real-time data and adjust pump operation remotely, enabling rapid response to changing conditions and reducing the need for on-site interventions.
5. **Performance Evaluation:** The project incorporated performance evaluation metrics to assess the effectiveness of the automated pump control system. Key performance indicators, such as energy savings, pump efficiency, and system reliability, were monitored to measure the impact of the implemented technologies.

Implementing automated pump control in the Netherlands resulted in several benefits, including improved energy efficiency, reduced operational costs, optimised pump performance, and enhanced system reliability. By leveraging intelligent computing and automation technologies, the project demonstrated the potential for optimising water supply network management and achieving sustainable operation.

7.8.3 Applications of Intelligent Computing and Automation to Integrated Decision Support Systems in Singapore

Implementing an integrated decision support system in Singapore's water supply network management is an exemplary case study that showcases the application of intelligent computing and automation technologies. The system utilised real-time sensor data, predictive analytics, and optimisation algorithms to support decision-making related to water allocation, demand forecasting, and infrastructure planning (Allen et al., 2012). The system enhanced operational efficiency, optimised resource utilisation, and enabled proactive response to changing water demand patterns. The project aimed to enhance decision-making processes, optimise resource allocation, and improve water management.

1. **Real-Time Data Integration:** The decision support system integrated real-time data from various sources, including sensors, weather forecasts, and water quality monitoring stations. This comprehensive data integration provided a holistic view of the water supply network's current conditions and enabled timely decision-making.
2. **Predictive Analytics:** Advanced predictive analytics techniques were employed to forecast water demand, predict water quality issues, and anticipate system performance under different scenarios. The system could generate accurate predictions and support proactive decision-making by analysing historical data and considering relevant factors.
3. **Optimisation Algorithms:** Optimisation algorithms were utilised to optimise resource allocation and operational planning. These algorithms considered water availability, demand patterns, energy consumption, and infrastructure constraints to generate optimal solutions. The system could recommend optimal pumping schedules, reservoir management, and water allocation.
4. **Scenario Analysis:** The decision support system facilitated scenario analysis by simulating different what-if scenarios to evaluate their potential

impact on the water supply network. This capability allowed decision-makers to assess the consequences of various decisions and policies before implementation, enabling informed decision-making and risk mitigation.

5. **Visualisation and Reporting:** The system provided visualisations and reports to present the analysed data and insights in a user-friendly manner. Graphical representations, dashboards, and reports helped decision-makers understand complex information and trends, facilitating effective communication and collaboration.

Implementing Singapore's integrated decision support system significantly improved water supply network management. The system enabled more informed and data-driven decision-making, optimised resource allocation, enhanced operational efficiency, and supported proactive responses to potential challenges. The project demonstrated the potential of intelligent computing and automation technologies in improving water management practices (Allen et al., 2012).

These case studies demonstrate the successful application of intelligent computing and automation in water supply network management (Adedeji et al., 2022;Lowe et al., 2022). Implementing these technologies has improved operational efficiency, optimised resource utilisation, reduced water losses, and enhanced decision-making capabilities (Mounce et al., 2010). These examples showcase the potential of intelligent computing and automation to transform water supply management and promote sustainability in diverse contexts.

7.9 CHALLENGES AND PROGRESS

7.9.1 Data Availability and Quality

The successful implementation of advanced control systems relies on accurate and timely data from sensors and monitoring devices. However, data availability and quality can be a challenge in water systems, as some areas may lack adequate sensor infrastructure or face data reliability and consistency issues. Efforts have been made to improve data availability and quality. Advances in sensor technology, data acquisition systems, and communication networks have led to increased sensor deployment and improved data collection. Additionally, data validation and quality assurance techniques, such as outlier detection and data fusion algorithms, have been developed to enhance the reliability and accuracy of collected data (Chandola et al., 2009).

7.9.2 System Complexity and Dynamics

Water systems exhibit complex dynamics, nonlinear behaviours, and uncertainties due to changing demand patterns, varying water quality, and hydraulic interactions. Designing control algorithms that handle these complexities and uncertainties is a significant challenge. Researchers have made progress in developing advanced control algorithms that can handle system complexity and dynamics (Wang et al., 2019). MPC, adaptive control, and optimisation techniques have been applied to water systems to effectively manage and control complex processes. These approaches consider system dynamics, constraints, and uncertainties to optimise control actions and improve system performance.

7.9.3 Implementation and Integration

Implementing advanced control systems in existing water infrastructure can be challenging due to limited resources, operational constraints, and compatibility issues with legacy systems. Integration of control systems with existing infrastructure and software platforms can also be complex. Efforts have been made to address these challenges through pilot projects, case studies, and collaboration between researchers, water utilities, and technology providers. Control system hardware and software advances have facilitated easier integration with existing systems. Additionally, developing standards and protocols for data exchange and interoperability has improved the compatibility and seamless integration of control systems in water infrastructure (Ulanicki & Ulanicka, 2017).

7.9.4 Human Factors and Operator Training

The successful implementation of advanced control systems requires trained operators and staff who can understand and effectively utilise the technology. Human factors, such as resistance to change, lack of awareness, and limited technical expertise, can pose challenges in adopting and utilising advanced control technologies. Efforts have been made to address human factors and operator training. Training programmes and capacity-building initiatives have been conducted to educate operators and staff about advanced control system technologies. Collaborative partnerships between researchers, technology providers, and water utilities have been established to facilitate knowledge transfer and support the implementation and operation of advanced control systems (Schmidt et al., 2024).

7.9.5 Cost and Return on Investment

The cost of implementing advanced control systems in water systems can be a barrier to their widespread adoption. Water utilities may face budget constraints and challenges in demonstrating the return on investment for implementing these technologies (Reis et al., 2023). The decreasing cost of sensors, control hardware, and data communication technologies has made advanced control systems more affordable. Additionally, studies and case examples demonstrate that the economic benefits, energy savings, and improved operational efficiency resulting from implementing advanced control systems have helped utilities justify the investment and realise the long-term benefits (Reis et al., 2023).

7.10 SUMMARY AND CONCLUSION

In this chapter, we discussed the topic of intelligent computing and automation in water supply network management. We explored various aspects of this field, including computational methods, big data analysis, water supply network data management, data compression, data aggregation, advanced control system technologies, adaptive control technologies, MPC, sensor integration, and their applications in water systems. The application of intelligent computing and automation technologies in water supply network management offers significant benefits, such as improved operational efficiency, proactive maintenance, and informed decision-making. However, the implementation of these technologies faces several challenges, including data management and integration, system complexity and dynamics, cybersecurity, human-technology interaction, and cost considerations.

Progress has been made in addressing these challenges. Data management techniques, modelling algorithms, and control strategies have improved the handling and integration of heterogeneous data sources and the modelling of complex system dynamics. Efforts to enhance cybersecurity and resilience have led to the implementation of secure communication protocols and training programmes. Human-centred design principles and training initiatives have improved user acceptance and effective human-technology interaction. The decreasing cost of computing hardware and technology, coupled with studies demonstrating economic benefits, has helped overcome cost barriers by making their implementation more affordable.

Intelligent computing and automation technologies have the potential to revolutionise water supply network management by enabling efficient operation, optimised resource utilisation, and proactive decision-making. While challenges exist, significant progress has been made in addressing them. Ongoing research, technological advancements, collaboration between stakeholders, and knowledge-sharing efforts are driving the adoption and implementation of these technologies in water systems. Continued integration of intelligent computing and automation will lead to more efficient and sustainable water supply network management, improved water quality, reduced energy consumption, and enhanced system resilience. Water utilities, researchers, and technology providers must work together to overcome the remaining challenges and leverage the full potential of intelligent computing and automation in water supply network management. By doing so, we can ensure the availability of clean and reliable water resources for the sustainable development of communities and the protection of the environment.

REFERENCES

Adedeji, K. B., Ponnle, A. A., Abu-Mahfouz, A. M., & Kurien, A. M. (2022). Towards digitalization of water supply systems for sustainable smart city development—Water 4.0. *Applied Sciences, 12*(18), 9174. https://doi.org/10.3390/app12189174

Agarwal, R., et al. (2019). Intelligent computing techniques for hydroelectric power generation and management: A review. *Renewable and Sustainable Energy Reviews*, *99*, 222–234.

Allen, M., Preis, A., Iqbal, M., & Whittle, A. J. (2012). Case study: A smart water grid in Singapore. *Water Practice and Technology*, 7(4), 2012089.

Astrom, K. J., & Wittenmark, B. (2013). *Adaptive Control* (2nd ed.). Dover Publications.

Bakker, M., Vreeburg, J. H. G., Palmen, L. J., Sperber, V., Bakker, G., & Rietveld, L. C. (2013). Better water quality and higher energy efficiency by using model predictive flow control at water supply systems. *Journal of Water Supply: Research and Technology—AQUA, 62*(1), 1–13.

Camacho, E. F., & Bordons, C. (2007). *Model Predictive Control*. Springer Science & Business Media.

Chandola, V., Banerjee, A., & Kumar, V. (2009). Anomaly detection: A survey. *ACM Computing Surveys (CSUR)*, *41*(3), 1–58.

Chen, X., & Wang, Y. (2018). Sensor technologies for water leak detection in pipe networks: A review. *Water*, *10*(6), 731.

Gouthaman, J., Bharathwajanprabhu, R., & Srikanth, A. (2011). Automated urban drinking water supply control and water theft identification system. In *IEEE Technology Students' Symposium* (pp. 87–91). IEEE.

Ioannou, P. A., & Sun, J. (2012). *Robust Adaptive Control*. Dover Publications.

Julier, S., & Uhlmann, J. (2004). Unscented filtering and nonlinear estimation. *Proceedings of*

Korbicz, J., Kowalczuk, Z., & Cholewa, W. (2011). *Fault Diagnosis: Models, Artificial Intelligence, Applications*. Springer.

Liu, H., & Xie, L. (2011). Fusion of multisensor data using Dempster-Shafer theory and belief functions. *IEEE Transactions on Systems, Man, and Cybernetics, Part A: Systems and Humans*, *41*(1), 182–201.

Lowe, M., Qin, R., & Mao, X. (2022). A review on machine learning, artificial intelligence, and smart technology in water treatment and monitoring. *Water*, *14*(9), 1384.

Luyben, W. L. (2013). *Control of Continuous Processes with Distributed Control Systems* (2nd ed.). Springer.

Maciejowski, J. M. (2002). *Predictive Control with Constraints*. Pearson Education.

Maksimovic, C., & Teixeira, A. P. (2017). *Sensor Technologies for Flood Monitoring and Management*. IWA Publishing.

Mohammadi, M. (2018). *Sensor Technologies for Water Quality Monitoring and Management*. IWA Publishing.

Mounce, S. R., Boxall, J. B., & Machell, J. (2010). Development and verification of an online artificial intelligence system for detection of bursts and other abnormal flows. *Journal of Water Resources Planning and Management*, *136*(3), 309–318.

Olsson, G. (2021). Urban water supply automation–today and tomorrow. *AQUA—Water Infrastructure, Ecosystems and Society*, *70*(4), 420–437.

Panchal, A., Dagade, K., Tamhane, S., Pawar, K., & Ghadge, P. (2014). Automated water supply system and water theft identification using PLC and SCADA. *International Journal of Engineering Research and Applications*, *4*(4), 67–69.

Perez, R., Sanz, G., Puig, V., Quevedo, J., Escofet, M. A. C., Nejjari, F., Meseguer, J., Cembrano, G., Tur, J. M. M. & Sarrate, R. (2014). Leak localization in water networks: A model-based methodology using pressure sensors applied to a real network in Barcelona [applications of control]. *IEEE Control Systems Magazine, 34*(4), 24–36.

Reis, A. L., Lopes, M. A., Andrade-Campos, A. & Antunes, C. H. (2023). A review of operational control strategies in water supply systems for energy and cost efficiency. *Renewable and Sustainable Energy Reviews, 175*, 113140.

Ross, T. J. (2010). *Fuzzy Logic with Engineering Applications* (3rd ed.). Wiley.

Savic, D., & Walters, G. A. (2015). *Advanced Control of Water and Wastewater Systems*. Butterworth-Heinemann.

Schmidt, J. Q., Roy, A., & Kerkez, B. (2024). "You can't write down the logic": Bringing smart technology into the water infrastructure control room. In *Proceedings of the 2024 CHI Conference on Human Factors in Computing Systems* (pp. 1–13).

Sutton, R. S., & Barto, A. G. (2018). *Reinforcement Learning: An Introduction* (2nd ed.). MIT Press.

Tong, Z., Ye, F., Yan, M., Liu, H., & Basodi, S. (2021). A survey on algorithms for intelligent computing and smart city applications. *Big Data Mining and Analytics*, *4*(3), 155–172.

Ulanicki, B., & Ulanicka, D. (2017). Challenges in the implementation of advanced control systems in water networks. *Environmental Engineering and Management Journal, 16*(5), 1053–1060.

Wang, Y., Puig, V., & Cembrano, G. (2017). Non-linear economic model predictive control of water distribution networks. *Journal of Process Control, 56*, 23–34.

Vilarinho, H., & Camanho, A. S. (2025). Challenges and trends in water system management. Natural resource management and policy. In J. C. Correia Leitão, W. Khairy, D. Pereira, J. Zascerinska, & P. D. Gaspar (Eds.), *Sustainable Water Management and Irrigation Systems: Tackling Climate Change* (pp. 71–90). Springer.

Yuan, Z., Olsson, G., Cardell-Oliver, R., van Schagen, K., Marchi, A., Deletic, A., … & Jiang, G. (2019). Sweating the assets—The role of instrumentation, control and automation in urban water systems. *Water Research, 155*, 381–402.

8 Wastewater Treatment in Water 4.0

8.1 INTRODUCTION TO SMART WASTEWATER MANAGEMENT

Water scarcity constitutes a worldwide challenge. A World Bank report cautions that water scarcity and climate change may result in losses of up to 6% of GDP for numerous countries and incite conflict. The United Nations World Water Development report indicates that roughly 80% of global wastewater is discharged into the environment without adequate treatment, despite its potential for recycling and reuse. UN estimates indicate that approximately 359 billion cubic meters of wastewater are generated annually globally. This quantity is approximately equivalent to 146 Olympic-sized swimming pools draining every minute. It is alarming to discover that over 2 billion individuals globally lack access to sufficient sanitation facilities, resulting in the discharge of untreated wastewater into lakes, rivers, and oceans. Pathogens, chemicals, and organic debris constitute perilous contaminants in this untreated wastewater (Tariq & Mushtaq, 2023). This necessitates an innovative methodology for wastewater treatment.

Wastewater is water that has undergone alterations in its physical, chemical, or biological properties due to the introduction of specific substances, making it unsuitable for use, including consumption. Human daily activities predominantly rely on water, resulting in waste discharge into aquatic environments. Included substances are bodily waste (faeces and urine), hair shampoo, hair, food remnants, fat, laundry detergent, fabric softeners, toilet paper, chemicals, household cleaners, soil, and microorganisms (germs), which can adversely affect human health and harm the environment. Much of the supplied water becomes wastewater, rendering its treatment essential. Water pollution is exceptionally challenging to remediate. Natural water filtration processes may require years or even centuries to remove all harmful contaminants, and even the most expensive modern treatments can take years. Two categories of pollutants require management. Water may be polluted by insecticides, herbicides, food processing byproducts, toxins from livestock operations, volatile organic compounds, heavy metals, chemical waste, and other substances. Nutrient pollution, including nitrogen and phosphates, leads to a proliferation of toxic algae that can be ingested by other aquatic species, resulting in lethality; it may also trigger outbreaks of fish diseases.

The source of water pollution must be eliminated to prevent further contamination. Pesticides, cleaning agents, and household chemicals may not pose a significant risk. Nonetheless, numerous chemicals, such as bleach, paints, thinner paint, and ammonia, are emerging as important concerns. Organic foods may be processed with chemicals; however, they are typically produced with minimal synthetic substances. Utilising organic foods diminishes the extent of chemical contamination

DOI: 10.1201/9781003633839-8

that affects the water supply. Municipalities can improve their wastewater treatment facilities to mitigate water pollution. The toxic substances generated in wastewater streams adversely affect human life, making separating and purifying water and wastewater streams crucial for numerous countries worldwide.

Moreover, toxic substances produced by chemical operations can cause catastrophic harm to water bodies. The ecotoxicological effects of textile wastewater containing dyes on the environment and living things are quite harmful. When azo dyes are present in the water, they prevent light from penetrating, impacting the growth of algae and aquatic plants. The absorption and reflection of sunlight by dying into water is a serious environmental issue. Light absorption has a significant influence on the food chain because it reduces the photosynthetic activity of algae, which impacts all creatures above them. After all, algae are at the bottom of the food chain. To remove pollutants from water and wastewater streams, efficient processes and techniques must be applied from an environmental standpoint.

Depending on the water quality and the region, many contaminants can be found in water, wastewater streams, and effluents. For instance, according to recent data from the World Bank, India has 18% of the world's population but only 4% of the water resources, making it one of the most water-stressed countries in the world. Reports predicted that 21 Indian cities—including Delhi, Bengaluru, Hyderabad, Mumbai, and Chennai, the most significant urban centres in India—will run out of groundwater by 2030, affecting 100 million people and leaving 40% of India's population without access to drinking water. Water shortage will impact India's population and the national GDP. Organic materials, pharmaceuticals and dyes, greasy waste, and heavy metals are all pollutants. Thin-film nanocomposite (TFN) membranes are extensively used in water treatment processes like reverse osmosis (RO) and nanofiltration (NF) processes because they provide high flux as well as rejection and high mechanical stability.

8.2 WASTEWATER TREATMENT PROCESSES AND THEIR EVOLUTION

Growing populations, heightened industrialisation, urban expansion, intensive agricultural practices, and various human activities have led to wastewater production, which has rendered water not only contaminated but also hazardous. Annually, millions of individuals succumb to illnesses transmitted via the ingestion of water tainted by harmful pathogens. This section examines the evolution of wastewater management over time. Wastewater treatment plants (WWTPs) are crucial for mitigating environmental pollution by treating and purifying wastewater before its discharge into the environment (Al-Hazmi et al., 2023). Nonetheless, the perception that these facilities are merely a sanitation infrastructure component is evolving. WWTPs are increasingly recognised as vital components of resource recovery, energy generation, and sustainable practices within a circular economy (Guven et al., 2023). This alteration signifies a fundamental transformation from perceiving wastewater as a burden to recognising it as a potential source of valuable resources. Historically, the primary objective of wastewater treatment has been to eliminate pollutants from water prior to its reintroduction into natural ecosystems, thereby safeguarding human health and the environment (Parveen & Khan, 2023). The "circular economy" concept has recently gained traction, advocating for the continuous circulation of resources within a closed-loop system to minimise waste and enhance resource efficiency (Kara et al., 2022). In this context, WWTPs are essential for implementing advanced strategies and technologies to recover valuable materials, generate renewable energy, and foster a more resilient and sustainable society.

According to Roth and Worthington (2023), traditional practices have been used before, including primary and secondary treatment. After wastewater enters the treatment facility, it passes through several steps to eliminate solid matter and big particles. To get rid of sand and gravel, grit removal and screening are required to eradicate big particles like debris. Following the screening, wastewater is sent to sedimentation tanks where heavier materials, such as oils and grease, sink to the bottom as sludge while lighter solids, such as grit, float to the top as scum. Water gets clarified due to this separation process, but pollutants are still suspended and dissolved. The primary treatment goal is the initial physical separation of the particles and liquids in the wastewater.

In the secondary treatment phase, microorganisms break down the wastewater's biodegradable organic pollutants. Creating an activated sludge involves mixing wastewater with a culture of microorganisms in aeration tanks. Aeration promotes the growth and degradation of contaminants by the microbes, which eat organic stuff. Pollutants are also broken down when wastewater flows down a bed of pebbles or synthetic media by microbial populations bonded to the medium. This process is known as trickling filtering.

8.3 INTEGRATION OF SMART TECHNOLOGY IN WASTEWATER MANAGEMENT: IOT, SENSORS, AND REAL-TIME MONITORING

The Internet of Things (IoT) and artificial intelligence (AI) have emerged as a game changer in the search for sustainable and efficient water management (Dada et al., 2024). When integrated, these innovative technologies form the backbone of smart water management systems (WMSs). The combination of IoT and AI enhances monitoring and control of water infrastructure. It provides decision-makers with powerful analytics and forecasting capabilities. The foundation of smart water management combines two critical technologies: IoT and AI. IoT refers to a network of interconnected physical items with sensors, actuators, and communication capabilities. These IoT devices cover the water management supply chain, from extraction sites to distribution networks and consumer endpoints. They collect real-time data on water quality, quantity, pressure, and other relevant properties, laying the groundwork for educated decisions.

In contrast, AI provides smart water systems with advanced analytics and decision-making capabilities. Machine learning (ML) algorithms within AI systems examine large datasets generated by IoT devices. They discover trends, predict, and turn raw data into actionable insights. This collaboration enables smart WMS to improve operations, predict future issues, and adapt to changing environmental conditions. IoT technologies are critical for better monitoring and control of water infrastructure. Intelligent sensors in water treatment plants constantly monitor essential properties such as turbidity, pH levels, and chemical concentrations. IoT-enabled devices across the distribution network provide real-time information on water flow, pressure, and leak detection. Smart meters with IoT capabilities precisely monitor water consumption, facilitating efficient billing and encouraging water conservation. The seamless integration of IoT into water infrastructure improves the precision and speed of data collection while allowing for remote monitoring and control. This capability is critical for the early detection of problems, preventative maintenance, and rapid emergency response, hence increasing the overall resilience of the WMS. AI improves the efficacy of smart WMS, particularly in optimising water treatment processes. AI systems using ML algorithms can analyse historical data to discover trends and patterns and anticipate future fluctuations in water quality. AI may help water treatment by optimising chemical dosages, energy usage, and operational efficiency. Adaptive learning algorithms continuously improve the system's performance using real-time feedback, supporting a more sustainable and adaptable water treatment process.

Furthermore, AI-driven predictive modelling can foresee equipment malfunctions, reducing downtime and maintenance costs. AI systems may identify irregular water quality or system behaviour and alert operators to potential problems before they escalate. This proactive method increases the reliability and efficiency of water treatment systems. The full potential of smart water management is reached through the seamless integration of IoT and AI technologies, which results in a complete and adaptive system. The permanent data streams from IoT devices serve as inputs for AI algorithms, which generate insights and actionable suggestions. This closed-loop communication enables real-time decision-making, allowing WMS to respond flexibly to changing conditions. For example, when IoT sensors detect a problem with water quality, AI algorithms may take automatic measures, such as altering treatment operations or contacting operators to request human assistance. The combination of IoT and AI transforms water management into a proactive, data-driven, and intelligent process that overcomes hurdles with agility and precision.

The integration of IoT and AI technology signifies a transformative change in water management. Intelligent WMS, supported by the synergistic application of various contemporary technologies, surpass conventional methods by providing a proactive and adaptive framework. These technologies empower decision-makers to address the intricacies of water management with unparalleled efficiency by delivering real-time data, optimising operations, and forecasting potential challenges. Integrating IoT and AI in the digital era offers a pathway to a more sustainable and resilient water future.

8.3.1 IoT-Water Management System (IoT-WMS)

Wastewater management systems are conceived explicitly to purify wastewater and efficiently provide clean water. However, this is a complex task due to frequent and unpredictable fluctuations of inlet wastewater flows arising from (random) rainwater or (periodical, e.g., day-night) sewage sources. Sometimes, these fluctuations also

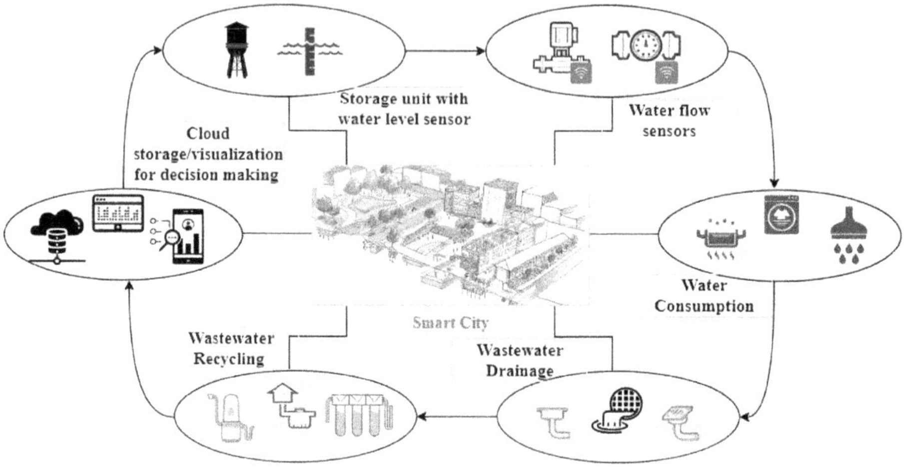

FIGURE 8.1 Overview of an integrated smart-city water management system, illustrating the flow of information and resources across key components, including storage units with water-level sensors, water-flow monitoring, consumption tracking, wastewater drainage, recycling processes, and cloud-based data visualisation to support informed decision-making and efficient resource management.

lead to failures and outages. Water scarcity is a serious worry as the world's population grows. This sparked the idea of wastewater reclamation and recycling. Sensors can be used for processing and monitoring throughout several stages of wastewater treatment. The IoT-WMS emphasises cloud security using blockchain technology to provide improved wastewater management systems in smart cities. This IoT-WMS proposes a trading system that uses blockchain incentives to promote wastewater recycling. Each residence and business in the smart city employs an IoT-based wastewater management strategy that includes sensors and actuators. Figure 8.1 depicts the basic conceptual framework of IoT in wastewater management for smart cities.

Figure 8.1 depicts the wastewater management cycle, which employs effective cloud data visualisation for informed decision-making. The decision support system distributes tokens (cryptocurrency) to homes and businesses based on the amount of recovered wastewater. To achieve a minimum volume requirement, consumers and companies resell these tokens. The token exchange parameters are established in a smart contract. Data kept on a blockchain-based cloud allows for immutable auditing. Smart contracts usually include an automatic tracking system hosted in a secure cloud environment. Supervised and unsupervised learning approaches detect people manipulating wastewater by recycling data in IoT meters, ensuring the model's durability. Efficient home automation maximises water use, hence improving the performance of the water supply system and services. This paper detects anomalies using a polynomial regression analysis technique. This anomaly detection technology was developed to identify theft of energy consumption power meters. This study explores a sequential learning model comprising the time-series data for meter readings. IoT enables knowledge acquisition and informed decision-making by gathering sensor data, such as soil moisture levels.

8.3.2 Water Management with Blockchain Technology

Figure 8.2 depicts the five tiers of an intelligent wastewater management system based on blockchain technology. The sensor layer is the first layer, and many IoT sensors are implemented to monitor water usage and wastewater reclamation. The secondary data collection layer reviews the data obtained from these sensors (see Figure 8.2). This layer includes various industrial and household infrastructures from layer 1, such as pipes with level gauges, water meters connected to storage tanks, and smart wastewater treatment units coupled with IoT-enabled devices. These systems can detect, monitor, analyse, gather, and interact with data on water storage levels, water consumption, wastewater production volumes, and recycling metrics. The data collected is then routed to the third layer. This edge computing layer is equipped with edge nodes/smart gateways. It uses Internet services such as WiFi, 4G, and 5G for processing, decision-making, collaborative filtering, and temporary data storage for a set period (Alzahrani et al., 2023).

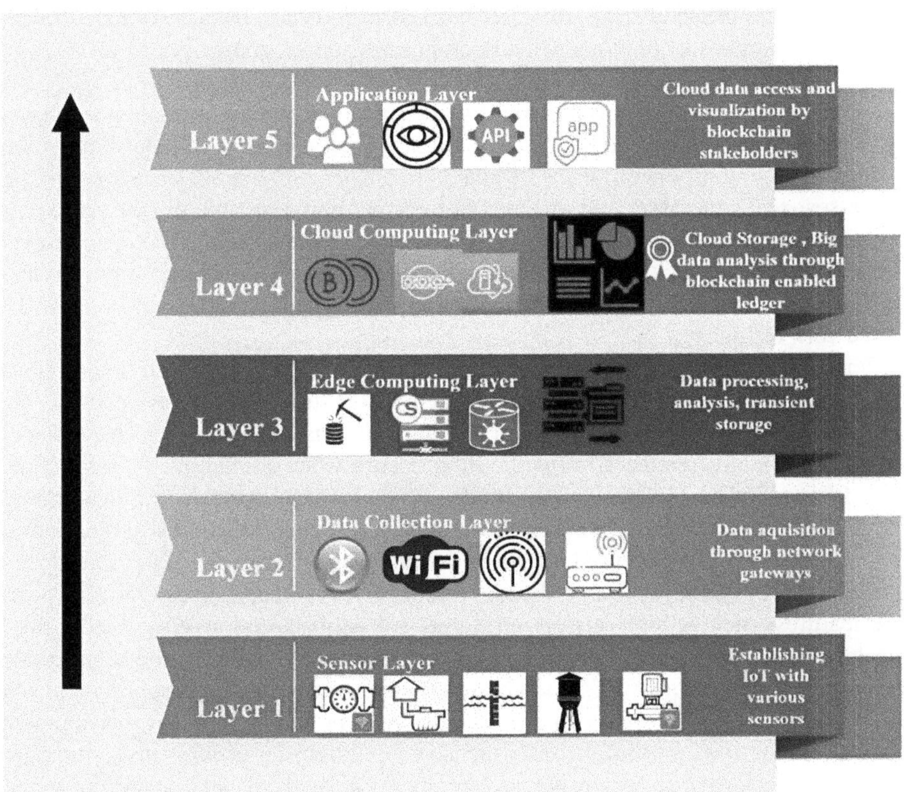

FIGURE 8.2 Five-layer architecture supporting a blockchain-integrated IoT framework. The layers progress from sensor deployment (Layer 1) and data acquisition through network gateways (Layer 2) to edge-level processing and temporary storage (Layer 3). Cloud-based big data analytics and blockchain-enabled storage form Layer 4, while Layer 5 provides application-level access, visualisation, and interaction for end-users and stakeholders.

The edge nodes use smart contracts to assess the acquired data items before incorporating them into the blockchain. The edge nodes transfer the validated data units to the cloud server in the fourth layer: the wastewater blockchain, which is located on cloud-based servers tasked with collecting, storing, processing, monitoring, and managing security operations enhanced by blockchain technology, which handles large amounts of data generated from various IoT-enabled smart cities. The fifth layer includes frameworks for managing and monitoring wastewater treatment for recycling and reuse. Smart contracts allow all stakeholders in smart city wastewater management and monitoring to query data stored in a blockchain-enabled cloud. Data can be accessed and analysed by authorised personnel for decision-making reasons.

8.4 TECHNOLOGIES AND INNOVATIONS IN WASTEWATER TREATMENT

The Fourth Industrial Revolution has greatly impacted academia, government policymaking, and the industrial sector. Regardless of industry, businesses must transition to cloud-based technology. Cloud computing enables companies of all sizes to create and adapt to the changes more quickly, providing them access to all of the benefits that Industry 4.0 will provide. By utilising cloud-based technologies, organisations may harness the power of technologically advanced applications and solutions to expand their operations. The goal is to enable autonomous decision-making processes, real-time asset and process monitoring, and real-time linked value creation platforms through early stakeholder engagement and vertical and horizontal connectivity. Many technologies are related to the fabrication of nanocomposites and thin-film membranes for water treatment and wastewater cleaning purposes in the framework of Industry 4.0.

The RO, NF, and membrane filtration (MF) technologies are intended to remove solute-dissolved substances from a pretreated liquid stream. Well-designed RO/NF systems are compact and need minimal maintenance, making them a feasible replacement for standard treatment trains. Fouling occurs when contaminants accumulate on a filter membrane's surface and hinder water passage through the membrane's pores. Various parameters like thickness, hydrophilicity, porosity distribution, density, and surface zeta potential are altered when such nanoparticles are introduced into the Polyamide layer. Using the right nanoparticles, membranes can gain positive features, such as high adsorption, improved antibacterial activity, higher photocatalytic activity, and good antifouling properties. Discovering the appropriate production parameters for TFN membranes that enhance water permeability and salt rejection is critical before they can be used extensively. TFN membrane development entails several time-consuming and expensive laboratory procedures involving poor separation performance, extensive surface chemical analysis, and degrading membrane stability. Estimating membrane performance with time is required for process analysis and optimisation and for designing new membrane processes. AI has recently been employed and shown great accuracy for better membrane separation, adsorption, higher fluid dynamic reaction, and separation applications. Artificial neural networks (ANNs) are the most prevalent AI modelling approach, and they have been widely utilised to simulate chemical and physical processes.

NF is associated with RO and ultrafiltration (UF). It is regarded as valuable and effective for removing heavy metal ions, including copper, nickel, and chromium, from water to facilitate its treatment for reuse (Raouf et al., 2019). Electrodialysis is a significant method for wastewater treatment (Mahmood et al. 2026). Only specific colloidal or dissolved molecules are exchanged through a selective ion exchange membrane between two liquids. Some charged entities can traverse the semi-permeable membrane, whereas oppositely charged entities cannot (Mahmood et al., 2026). The membrane is a barrier that divides into two phases of natural or synthetic materials, including polymers, ceramics, or metals. The two fundamental types of membranes are cation-exchange and anion-exchange membranes. This technique has been extensively employed to purify industrial effluents, recover valuable elements from these effluents, produce salt, and generate potable and processing water from brackish and saline sources (Raouf et al., 2019). Membrane technology was initially called "Desalination technology" (Fane et al. 2010). Oily contaminants can be efficiently eliminated by integrating various technologies, including photocatalytic reactors with UF for oil bilge effluent and employing RO with adsorbents for water produced in oilfields (Moslehyani et al., 2016).

8.5 AUTOMATED AND AI-DRIVEN CONTROL SYSTEMS FOR PROCESS OPTIMISATION

The adsorption equilibrium concentration of pollutants and the adsorption capacity of the adsorbent as more than just a function of adsorption variables may be simulated through AI technology, such as solution pH or a solid dose, solute starting concentration, and medium temperature. The optimal weights for an optimisation strategy are used to find an ANN model that minimises a performance index during training. The universal approximation theorem states that a multiple-layer-based feedforward neural network framework integrates a single hidden layer and can approximate any continuous and nonlinear mapping function. The number of neurons in the hidden layer with initial weights and biases, the optimisation strategy, and other parameters should be chosen carefully because they significantly impact the ANN models' working. ML approaches are generally helpful tools increasingly employed in several complex systems, given the enormous number of factors and the complexity of membrane processes. ANNs have been used to manufacture and separate various membranes, including NF, UF, RO, and MF. Using ML approaches, nonlinear functions can detect complex correlations between input and output data. ANNs are increasingly used in water treatment applications due to their ability to represent complex processes. This work proposed an approach for developing and verifying ANNs for applications requiring drinking water quality and water distribution systems (WDS). ANNs are not dependent on biological, pharmacological, or physical processes. Instead, these empirical models recognise that, given specific inputs, some processing conditions may be modified to achieve a specific goal based on "learning" from earlier training. However, the performance of the ML method's estimation and generalisation capabilities depends on the training method and architecture used. Through a training phase, ANN link input and output data were used to measure weights among neurons along with biases and linear and nonlinear functions.

8.6 INNOVATIONS IN TREATMENT TECHNOLOGIES: MEMBRANE BIOREACTORS, ADVANCED OXIDATION, AND NUTRIENT RECOVERY

Water is crucial for maintaining natural balance and human growth, but industrialisation, urbanisation, and agricultural expansion have led to organic pollution contamination in water bodies. Pollutants from industrial wastewater, agricultural runoff, urban sewage, and human activities pose toxicity, carcinogenicity, and mutagenic risks to water quality and the environment (Bukola et al., 2015; Madhav et al., 2020; Rathi et al., 2021). It then calls for an urgent need to investigate efficient, cost-effective, and ecologically acceptable ways of removing organic contaminants from water bodies. In recent decades, photocatalytic technology has gained significant interest as an environmentally beneficial solution to pollutant degradation. Because of their bandgap, semiconductors are being utilised in advanced oxidation processes (AOPs) to catalytically destroy organic contaminants, particularly those that can absorb visible light. Integrating N-TiO_2 with $ZMoS_2$ at varying energy levels improves the effectiveness of electron-hole separation and photocatalytic activity (El Sharkawy et al., 2023; Li et al., 2023).

The principle of photocatalytic degradation/reduction of inorganic pollutants is like that of photocatalytic degradation of organic pollutants in wastewater. It only concentrates on contaminants, such as heavy metals and various anions. Heavy metal ions, essential for metabolism, represent a danger to water resources owing to their high concentrations. The growth of metallurgy, mining, nuclear energy, and chemical manufacture generates significant amounts of harmful heavy metal ions, harming organic cells and creating health issues. These ions cannot be biodegraded; thus, they accumulate in people and animals via food chains and drinking water (Mishra et al., 2019; Vardhan et al., 2019; Khalef et al., 2022). As a result, it is critical to remove harmful heavy metals in sewage before infiltrating the ecosystem. Photocatalytic heavy metal ion removal in water involves converting hazardous high-valence ions into low-valence or zero-valence meta-ions.

An AOP has been applied to wastewater from different sources. For instance, an AOP was used by El-Gawad et al. (2023) to treat industrial crude oily wastewater containing greases, fats, and oils mixed with dissolved organic and inorganic substances suspended in high concentrations. The oil removal percentage was measured when H_2O_2 was added and not added. With 3 mL of H_2O_2, the percentage removal of oil was 99%, whereas when it was not added, the percentage removal was 75%, all in a reaction time of 40 minutes. This established that H_2O_2 can significantly improve the efficiency of AOPs. The study further ascertained that photocatalytic oxidation by H_2O_2/UV/TiO_2 can treat wastewater effectively. Another survey by Ghjair and Abbar (2023) confirmed that an AOP is effective in wastewater treatment when they attempted to remove chemical oxygen demand (COD) from wastewater by using Electro-Fenton (EF) and Solar Electro-Fenton (SEF) processes on wastewater samples obtained from Al-Diwaniyah Hospital in Iraq. The observed impact of H_2O_2/COD ratio, current density, and time on the treatment processes varied with different parameters. They found the efficiency of EF to depend on two factors: the H_2O_2/COD ratio and the current density. Still, the H_2O_2/COD ratio had the most significant effect

(96.23%) on removing COD in hospital wastewater, followed by the current density. On the other hand, COD removal shows a rate of 90% when SEF was adopted, fixing at a density of 10 mA/cm^2 and COD ratio of 0.2. The COD removal rate was reduced to 88.31%, especially with low energy consumption.

In their study, Ponnusami et al. (2023) also demonstrated the efficacy of integrating AOPs with biological processes for wastewater treatment. The review determined that applying AOPs to treat wastewater results in a quantifiable level of compliance and efficacy. In their study, Wang et al. (2023) conducted a comprehensive analysis of the utilisation of AOPs for the remediation of nano or microplastic contaminants. It has been found that AOPs have proven to be efficient in treating wastewater contaminated with micro/nanoplastics. Additional uses of AOPs can be observed in eliminating organic substances such as dyes, pharmaceuticals, pesticides, and inorganic pollutants like heavy metals and nitrates. AOPs are employed in the processing of both domestic and industrial wastewater, as well as in drinking water purification.

Mechanisms of AOPs: Deng and Zhao (2015) highlighted that AOPs function by generating hydroxyl radicals through following methods.

1. **Hydroxyl Radical-Based AOPs:** This includes processes like Fenton and photo-Fenton reactions, ozonation, and UV/H_2O_2 treatment. Hydroxyl radicals are highly reactive and can oxidise many organic pollutants.
2. **Ozone-Based AOPs:** Ozone alone or combined with UV light or hydrogen peroxide can generate hydroxyl radicals, enhancing oxidation.
3. **UV-Based AOPs:** Ultraviolet light produces hydroxyl radicals through photolysis when used with hydrogen peroxide or titanium dioxide.
4. **Fenton-Related AOPs:** These involve the reaction of hydrogen peroxide with iron salts to produce hydroxyl radicals. Variants like photo-Fenton use UV light to enhance the reaction.
5. **Sulphate Radical-Based AOPs:** Sulphate radicals (SO_4^-) have gained attention for their high oxidation potential and stability over a more comprehensive pH range than hydroxyl radicals. Dealing sulphate radicals includes activating persulphates or peroxymonosulphates using heat, UV light, or transition metals.
6. Other AOPs include electrochemical processes and ultrasonic irradiation, which can produce hydroxyl radicals.

AOPs have been applied to various wastewater samples, including landfill leachate, which explains that high-strength wastewater contains numerous organic contaminants. AOPs can effectively reduce the concentration of these pollutants, transforming them into less toxic forms (Deng & Zhao, 2015; Cheng et al., 2016; Sharma et al., 2018). Effluent, Organic Matter in Biologically Treated Secondary Effluent also explains that AOPs can mitigate UV-quenching organic matter, improving the quality of treated effluent for reuse (Deng & Zhao, 2015). The efficiency of AOPs depends on factors such as the type of wastewater, the concentration of pollutants, and the operational conditions. However, a few potential challenges include the short half-life of radicals, high operational costs, and the need for pre-treatment or post-treatment steps to handle byproducts or remaining contaminants.

Another emerging wastewater treatment method is MF. MF can be classified into RO, UF, NF, and electrodialysis, the parts that aid the treatment and are classified as their usage method. UF and microfiltration are popular membrane methods for removing particles from wastewater treatment systems. Membranes may be introduced to the biological wastewater treatment process separately or as part of the process. If the membranes are applied as a separate unit technique, they are often called the "polishing" phase and are generally based on the hollow fibre membrane design. Technologies such as RO and NF remove many contaminants, including dissolved solids and microorganisms. These devices are increasingly being employed for water reuse purposes. RO allows the treated water to pass through a semi-permeable membrane while blocking the contaminant (Raouf et al., 2019). The method greatly dissolves some particles and pollutants in water. On the other hand, UF is a process that treats the wastewater at a lower membrane rate of pressure for the ejection of colloidal and dissolved materials in the water (Pachwarya et al., 2021).

Chemical demulsification is another method of treating wastewater. The steps involved in demulsification are just as complex as those in emulsion creation. As previously stated, emulsions consist of a disseminated phase, a continuous phase (sometimes called the outside component), and a substance that emulsifies. They are stable and complicated mixtures. Therefore, for demulsification to be effective, it must be able to reduce or deactivate the phases' stability, causing them to separate (Medeiros et al., 2022). Three strategies create physical, chemical, and biological demulsification techniques. In certain instances, demulsification can also be broken down into three operational steps: oil aggregation, gravity separation, and destabilisation of the oil-in-water or water-in-oil interface. The physical approach, which often requires high energy consumption and expensive operation, involves separating the components by mechanical, thermal, or electrical action. During mechanical demulsification, forces break down the physical barrier and separate the components. Among the equipment used for the mechanical separation process are centrifuges, gravity-settling tanks, and cyclones (Medeiros et al., 2022). Another physical technique is electrical demulsification.

The four primary kinds of petroleum hydrocarbons in oily wastewater are aliphatic, aromatic, asphaltenes, and compounds containing nitrogen, sulphur, and oxygen. It comprises organometallic complexes, including vanadium, nickel, cadmium, and lead. Regarding the biological method, demulsification may be accomplished using pathogens and their byproducts with various biochemical structures. These organisms possess systems that allow them to access the emulsion's hydrocarbons for sustenance. By substituting extracellular amphiphilic biosurfactants, or their hydrophobic cell surfaces made of proteins and lipids, for emulsifying molecules, the demulsifying process lowers the tension that exists between the emulsion phases (Medeiros et al., 2022). Effective chemical demulsification techniques for separating emulsions rely on elements like chemical choice and concentration. The components are homogenised with the emulsion during the procedure to guarantee that the reaction occurs. The mixture must be heated to speed up the separation process or break the emulsion entirely. It is necessary to give the phases enough time to separate after these procedures (Medeiros et al., 2022).

8.7 CASE STUDIES SHOWCASING EFFICIENT WASTEWATER TREATMENT SOLUTIONS

8.7.1 Advanced Oxidation Processes and Recycled Water in the United States

The United States has invested in AOPs for wastewater treatment and recycled water programmes to support agricultural irrigation (Ameta & Ameta, 2018). AOPs involve using chemical oxidants, such as ozone or hydrogen peroxide, combined with UV light to break down pollutants in wastewater. Treated wastewater is recycled and distributed for agricultural use through dedicated pipelines. The Orange County Water District groundwater replenishment system treats wastewater using AOPs and provides high-quality recycled water for irrigation in Orange County, California. Monterey Regional Water Pollution Control Agency uses recycled water to irrigate artichokes, lettuce, and strawberries in the Salinas Valley. AOPs effectively remove a wide range of contaminants, ensuring the treated water is safe for agricultural use. Recycled water programmes enhance water security for farmers, particularly during droughts. Using recycled water reduces the strain on freshwater resources and promotes sustainable agricultural practices.

8.7.2 Membrane Bioreactors and Irrigation Schemes in Australia

Australia has adopted membrane bioreactor (MBR) technology for wastewater treatment, facilitating the reuse of treated water in agriculture. MBR technology combines biological treatment with MF to produce high-quality effluent (Asif et al., 2022).

Treated water is distributed to agricultural areas for irrigation purposes. Western Treatment Plant, Melbourne plant uses MBR technology to treat wastewater, providing high-quality recycled water for irrigation in the Werribee Irrigation District. Under the Virginia Pipeline Scheme, South Australia, recycled water from MBR-treated sewage is used to irrigate vineyards and horticultural crops. MBR technology produces water that meets stringent quality standards for agricultural use. The availability of reliable recycled water supports agricultural productivity and economic growth. It is, therefore, imperative to reduce the discharge of untreated or partially treated wastewater into the environment.

8.8 WATER REUSE STRATEGIES FOR AGRICULTURE, INDUSTRY, AND MUNICIPAL USE

Wastewater is now recognised as a valuable source of clean water for both non-potable and potable applications. Water is filtered and treated for human consumption and many other uses, such as industrial, irrigation, and washing. Recycled wastewater is utilised for a variety of applications and has the potential to deliver societal benefits in a variety of ways. Wastewater treatment is essential to increase clean water availability. Whereby clean water is a necessity for individuals today as there are various health issues related to wastewater and sewage. Other than personal adversities, clean water is a requirement for societies and growth opportunities. Water could be considered a considerable element for development to take place in society. Without water, it is challenging to incline towards any progress or growth, so it has become vital to ensure ample clean water is available for use. It then calls for clear goals and objectives for treating wastewater. Table 8.1 highlights the main objectives and goals of wastewater treatment. These objectives align with the global Sustainable Development Goals (SDGs) related to water availability, according to Obaideen et al. (2022) and Qasim & Zhu (2017).

Agriculture faces water security and sustainability challenges due to water shortages and pollution. Population growth and climate change have exacerbated water unpredictability, making sustainable water management techniques crucial. Wastewater recycling is a reliable option, as water is an essential component of everyday life for humans, animals, and plants. Recycling and reusing wastewater in agriculture can increase water availability for agricultural use, as wastewater has been proven to affect agricultural productivity significantly. Therefore, sustainable water management techniques are essential for the agriculture industry in the 21st century (Jaramillo & Restrepo, 2017).

Babayan et al. (2012) posited that most waste from sewage is rich in nutrients, resulting in increased productivity, soil fertility, and plant growth. Accurately refined wastewater could improve seed germination, crop yield, crop nutrient accumulation, and other important soil characteristics. However, this calls for a novel practice in wastewater treatment to enhance the preservation of these nutrients and encourage safe agricultural practices. Diverse technologies portraying innovative wastewater treatment and reuse have emerged as inevitable components in tackling these scarcity concerns. These innovations have provided answers for pollution reduction and resource recovery and strengthened agricultural systems' resilience to climatic unpredictability (Abonyi et al., 2024). Advanced technologies, such as MF, biological treatment, phytoremediation, and decentralised treatment systems, among others, have been used to convert wastewater into a valuable asset for agricultural output (Al Hamedi et al., 2023).

TABLE 8.1
Goals and Objectives of Wastewater Treatment

S/N	Goals and Objectives
1	To improve wastewater quality and reduce wastewater strength and volume.
2	Elimination of pollutants, toxicants, and many such (physical, chemical, and biological compounds).
3	Preservation of water quality of natural water resources and the conservation of aquatic ecosystems.
4	Making wastewater fit for reuse and usable for other purposes.
5	Prevention of harmful diseases and ensure the reduction of pathogens.
6	Attain Sustainable Development Goals (SDGs) standards.
7	Minimise the impacts of pollutants on the ecosystem and public health.
8	Ensure the protection of public safety and health.

These are three methods that professionals in the water and wastewater industries use to increase reuse, decrease consumption, and make facilities more efficient. These strategies may help companies reduce water consumption by 20–50%.

8.8.1 Closed-Loop Water Reclamation Systems

Closed-loop systems are designed to reclaim all the water they use. They are especially effective in facilities where equipment, vehicles, and other assets must be regularly washed, as they can streamline the recollection and reuse of wash water. For example, reuse—as opposed to single-use—clean-in-place (CIP) systems have long been popular in the food and beverage sector. These systems allow facilities to clean equipment to industry standards and federal regulations while improving the site's water efficiency. They clean equipment without requiring its disassembly.

In many systems, water from the final rinse is recirculated, carrying heat and cleaning chemicals that make it more effective than the pre-rinse of other on-site equipment. Captured water can also be used for clean-out-of-place operations that require machinery disassembly. By contrast, single-use CIP systems may capture used wash water but will not reuse it in cleaning. Instead, it will be immediately looped into the facility's wastewater treatment systems. Reuse CIP systems can make it easier to recapture and use wash water in any facility where equipment must be regularly cleaned. Similarly, finding ways to close the loop in existing systems can streamline and simplify industrial water usage.

8.8.2 Zero Liquid Discharge

Some facilities may want to adopt site-wide systems that aim to minimise or eliminate water use. Zero liquid discharge (ZLD) is an engineering approach to water treatment that seeks to recapture all site water and reduce contaminants to solid waste. Ideally, the practice eliminates liquid waste from leaving a plant, ensuring that water entering the facility remains in circulation for as long as possible. These systems are often considered an alternative or upgrade to more conventional on-site treatment strategies, like filtration pools, which may lose water due to evaporation or incomplete filtration of wastewater streams. The approach can be energy—and time-intensive in implementation, but advancing technology has made ZLD much more practical. For example, modern facilities striving for ZLD typically use RO filtering instead of thermal or evaporation filtration, which reduces the energy and money necessary to achieve ZLD.

8.8.3 Routine Maintenance and Equipment Upgrades

In addition to strategies that promote reuse, it is also important to remember how inefficient equipment and leaky pipes can contribute to facility consumption through unaccounted-for water. Ageing equipment and water infrastructure can significantly increase loss and waste in older facilities and places where routine maintenance has been neglected. Even equipment installed just ten years ago may be considerably less efficient than modern alternatives. For professionals working in older facilities, performing an audit to identify where the facility may be losing water is an effective strategy before recommending installing new systems and upgrades. Planning for leaks and equipment failure can help businesses stay on top of water consumption.

8.9 NUTRIENT AND ENERGY RECOVERY FROM WASTEWATER

Sniatala et al. (2023) listed various technologies that have been employed to treat wastewater by extracting resources from sewage, which include nutrient recovery, energy production from biogas, forward osmosis (FO), AOPs, and intelligent control and monitoring systems. Two important nutrients that can be found in wastewater are phosphorus and nitrogen. Utilising sophisticated techniques such as struvite precipitation or enhanced biological phosphorus removal, phosphorus can recover in the form of struvite, a fertiliser with a gradual release. Furthermore, methods for recovering nitrogen for use in agriculture in the form of ammonium or nitrate have been developed. Biogas (mainly methane) is produced by anaerobic digestion, a process in which microbes decompose organic matter without oxygen. This biogas can be collected and used as a clean energy source for heating, producing power, or fuel for cars. While Odirile et al. (2021) discussed renewable energy sources and waste management, they mentioned anaerobic digestion to create biogas from municipal sewage sludge.

Furthermore, FO is a newly developed membrane technology that extracts water from impurities using osmotic pressure. It can create high-quality water to be reused or concentrate wastewater streams for resource recovery. AOPs use strong oxidants, such as hydroxyl radicals, to degrade persistent organic contaminants in wastewater. These procedures successfully remove pharmaceuticals, personal hygiene items, and other newly discovered pollutants that are difficult to eradicate using traditional treatment techniques. WWTPs can optimise treatment operations using real-time monitoring systems and sensor technology. By giving operators access to data on variables like pH, dissolved oxygen, and nutrient levels, these systems allow for timely modifications that improve efficiency and maximise resource recovery.

Biogas is an environmentally friendly natural gas produced by anaerobic digestion of organic materials. It comprises trace amounts of carbon dioxide, methane, and other gases (Atelge et al., 2020). Biogas, an environmentally sustainable energy resource, is primarily produced by the anaerobic decomposition of organic matter, including agricultural byproducts, food remnants, sewage residue, aquatic sediments, animal faeces, crop waste, and wastewater. Biogas generation is also aided by human activity in places like landfills and water lagoons (Wid & Raudin, 2023). Biogas is produced using various methods, including landfill gas recovery systems, wastewater treatment facilities, and biodigesters (Nguyen et al., 2021). To produce biogas, feedstock availability and government policy assistance are essential. Generating biogas is paramount in resolving the escalating problems regarding energy security, environmental sustainability, and waste management. The proposed method presents a potential opportunity to mitigate greenhouse gas emissions, enhance energy source diversification, and foster the principles of a circular economy (Fajobi et al., 2022; Wang et al., 2023).

8.10 ENVIRONMENTAL AND REGULATORY CONSIDERATIONS FOR SAFE REUSE PRACTICES

Various ancient cultures have already developed sanitation infrastructure and wastewater management techniques. For example, in Ancient Rome, the Cloaca Maxima was constructed as an extensive sanitation and drainage system in response to flooding around the 6th century B.C. The quantity and makeup of wastewater generated by households, hospitals, and industries indicate an important matter for people's and the environment's health. Thus, effectively handling wastewater, which involves collection, treatment, and disposal with the most minor viable ecological impact, is necessary (Villarín & Merel, 2020; Saravanan et al., 2021). Wastewater management is still a concern to minimise the environmental effect on receiving waterways and safeguard public health. For example, continuous research strives to improve the attenuation of newly emerging pollutants produced by contemporary civilisations by developing novel treatment techniques (Massoud et al., 2009; Calvo-Flores et al., 2018; Mafuta et al., 2021).

Furthermore, this swift and recent change is also a consequence of wastewater management being prioritised more and more worldwide rather than being seen as a local problem that each municipality must deal with on its own. For instance, increased societal and political awareness has resulted from information about phenomena like the "feminisation" of fish populations downstream of water treatment facilities caused by endocrine disruptors discharged with wastewater discharges (Arlos et al., 2015; Hicks et al., 2017; Villarín & Merel, 2020). Research on wastewater is influenced by paradigm modifications brought about by the necessities of contemporary society, which also alter how wastewater is seen. For example, as populations in urban agglomerations rise, so does the water demand, worsening big cities' water footprints. This promotes the paradigm change that turns wastewater from an undesirable material into a priceless resource (Villarín & Merel, 2020; Giwa et al., 2023).

The use of the conventional (traditional) approach to wastewater treatment has been the primary focus of the existing WWTPs, and this approach comes with various challenges despite its advantages (Mo & Zhang, 2013; Cao et al., 2022; Xiao-jun et al., 2023). A significant challenge of the traditional method is the operational cost, which is relatively high due to the necessity for energy and chemical inputs, which could be costly. Another similar challenge is the limitation of the method for nutrient recovery, which is essential for the supply of treated wastewater, especially for agricultural and industrial use (Chrispim et al., 2019; Rosemarin et al., 2020). Though effective in wastewater treatment, the conventional method may not adequately remove pathogens, which may pose a serious health risk to the consumers of the treated water products using this method. Due to these limitations, there is a serious need for a paradigm shift in the wastewater treatment sector and the need to formulate an extensive policy, institutional and regulatory frameworks to accommodate the inclusion of novelty, innovative and sustainable technologies in the treatment of wastewater, which will ensure long-term environmental and public safety benefits.

While this work advocates a policy framework to guide the recycling (treatment), reuse, and resource recovery of wastewater and an institutional framework to ensure an all-inclusive implementation of the policy, there remains a need for a regulatory framework to provide a basis for ensuring the enforcement of the policies. A well-established regulatory framework will incentivise policy implementation (Rodriguez et al., 2020).

In addition to the policy, standards and guidelines should inform the operational modality of the policies. These standards and procedures must be based on internationally acceptable beliefs, be worth national prestige, and be carried out duly. To protect public health and the environment, especially to avoid waterborne infections and diseases, wastewater collection, treatments, and careful disposal become imperative. On the other hand, water resource recovery becomes essential to prevent unnecessary waste of water resources. Hence, formidable effluent discharge standards must serve as a template for developing nations in a collective effort to alleviate the challenges attached to improper wastewater disposal. Two common or conventional waste disposal methods are surface water discharge and in-land application systems. The surface discharges disposed of sewage into surface waterbodies like rivers, streams, and oceans. At the same time, the in-land systems drain the water into the groundwater, and sometimes, it sleeps into the surface water, especially when discharged close to a surface waterbody. These disposal methods determine the wastewater treatment and recycling approach to be adopted.

There are two equally effluent disposal/discharge standards that regulatory agencies often set: the first is the technology-based effluent discharge standard, and the second is the water quality-based limit (also known as in-stream or environmental standards). These waste discharge standards are enacted to safeguard water disposal and rightly channel the waste to the correct recycling destination. Developing a stereotype guideline for the safer use of wastewater is relatively tricky since peoples' actions and acceptance of various innovations, especially in Africa, are guided by factors like religion and sociocultural beliefs. In many parts of the continent, for instance, wastewater reuse, especially from faecal sludge, is considered disgusting and well-denied. In Asia and some other regions, wastewater and proceeds from faecal sources are well-recognised and considered safe (Mafuta et al., 2021). Before communities where Islam is their prevalent religion will allow for wastewater reuse, it must have been purified and diluted. Therefore, central water agencies must note that each country's water reuse policies must consider their country's culture and religion.

The regulatory framework should provide a standard for ensuring compliance and monitoring the implementation of policies guiding wastewater collection, treatment or nutrient recovery and reuse. A vast regulatory system must cover wastewater byproducts, industrial waste discharge, and intersectoral regulations. Effective monitoring needs to be established to track progress and avoid long-term deterioration. This monitoring exercise will be sustained if the knowledge that the wastewater problem is intergenerational is disseminated. The government should set up a monitoring and evaluation body, which could comprise government employees and members of the society; this body will be saddled with the responsibility of ensuring compliance in implementing the policies. Penalties like pollution tax and defaulters' fines must be made available for defaulters, and the monitoring team should be vested with the power to enforce the practice without prejudice. The bodies providing incentives for the treatment plants and resource retrieval firms must also ensure compliance. The monitoring team will oversee ensuring compliance with such bodies.

8.11 CHAPTER SUMMARY AND CONCLUSION

This chapter examines the evolution and innovations in wastewater treatment, emphasising its critical role in easing global water scarcity. It emphasises the critical need for proper wastewater treatment, particularly in water-scarce countries like India, where urban areas face imminent groundwater depletion. The investigation covers classic and modern wastewater treatment systems, including primary and secondary techniques, AOPs, and membrane technologies, including NF and RO. Smart technologies like IoT and AI transform wastewater treatment systems into intelligent, adaptable networks that increase operational efficiency, predictive maintenance, and resource recovery. Case studies in Australia and the United States demonstrate the practical application of recycled water in agriculture and the efficient implementation of MBRs. Nutrient and energy recovery, particularly biogas generation, is being prioritised to support circular economies. Furthermore, the chapter highlights the importance of legislative frameworks and cultural elements in ensuring the safe and sustainable reuse of wastewater globally. Efficient wastewater treatment is critical for environmental preservation and ensuring global water security and sustainability. Wastewater may be converted into a resource for clean water, power, and nutrients by moving away from traditional ways and towards innovative, data-driven solutions. This paradigm shift demands collaborative efforts, strong legal frameworks, and cultural adaptability to maximise the potential for wastewater reuse in agriculture, industry, and communities. Adopting intelligent technology, advanced treatment techniques and sustainable practices can help establish strong WMS that can address current and future challenges.

REFERENCES

Abonyi, M. N., Obi, C. C., Nwabanne, J. T., & Aniagor, C. O. (2024). Emerging and eco-friendly biological methods for agricultural wastewater treatment. *Environmental Systems Research, 13*(1), 45.

Al Hamedi, F. H., Kandhan, K., Liu, Y., Ren, M., Jaleel, A., & Alyafei, M. A. M. (2023). Wastewater irrigation: A promising way for future sustainable agriculture and food security in the United Arab Emirates. *Water, 15*(12), 2284.

Al-Hazmi, H. E., Mohammadi, A., Hejna, A., Majtacz, J., Esmaeili, A., Habibzadeh, S., Saeb, M. R., Badawi, M., Lima, E. C., & Mąkinia, J. (2023). Wastewater treatment for reuse in agriculture: Prospects and challenges. *Environmental Research, 236*, 116711.

Alzahrani, A. I., Chauhdary, S. H., & Alshdadi, A. A. (2023). Internet of things (IoT)-based wastewater management in smart cities. *Electronics, 12*(12), 2590.

Ameta, S. C., & Ameta, R. (Eds.). (2018). *Advanced oxidation processes for wastewater treatment: emerging green chemical technology*. Academic Press.

Arlos, M. J., Bragg, L., Parker, W., & Servos, M. (2015). Distribution of selected antiandrogens and pharmaceuticals in a highly impacted watershed. *Water Research, 72*, 40–50.

Asif, M. B., Zhang, Z., Vu, M. T., Mohammed, J. A., Pathak, N., Nghiem, L. D., & Nguyen, L. N. (2022). Membrane bioreactor for wastewater treatment: Current status, novel configurations and cost analysis. In *Cost-Efficient Wastewater Treatment Technologies: Engineered Systems* (pp. 147–167). Springer International Publishing.

Atelge, M., Krisa, D., Kumar, G., Eskicioglu, C., Nguyen, D. D., Chang, S. W., Atabani, A., Al-Muhtaseb, A. H., & Unalan, S. (2020). Biogas production from organic waste: Recent progress and perspectives. *Waste and Biomass Valorization*, *11*, 1019–1040.

Babayan, M., Javaheri, M., Tavassoli, A., & Esmaeilian, Y. (2012). Effects of using wastewater in agricultural production. *African Journal of Microbiology Research*, *6*(1), 1–9.

Bukola, D., Zaid, A., Olalekan, E., & Falilu, A. (2015). Consequences of anthropogenic activities on fish and the aquatic environment. *Poultry, Fisheries & Wildlife Sciences*, *3*(2), 1–12.

Calvo-Flores, F. G., Isac-García, J., & Dobado, J. A. (2018). *Emerging Pollutants: Origin, Structure, and Properties*. John Wiley & Sons.

Cao, M., Hu, A., Gad, M., Adyari, B., Qin, D., Zhang, L., Sun, Q., & Yu, C.-P. (2022). Domestic wastewater causes nitrate pollution in an agricultural watershed, China. *Science of the Total Environment*, *823*, 153680.

Cheng, M., Zeng, G., Huang, D., Lai, C., Xu, P., Zhang, C., & Liu, Y. (2016). Hydroxyl radicals based advanced oxidation processes (AOPs) for remediation of soils contaminated with organic compounds: A review. *Chemical Engineering Journal*, *284*, 582–598.

Chrispim, M. C., Scholz, M., & Nolasco, M. A. (2019). Phosphorus recovery from municipal wastewater treatment: Critical review of challenges and opportunities for developing countries. *Journal of Environmental Management*, *248*, 109268.

Dada, M. A., Majemite, M. T., Obaigbena, A., Daraojimba, O. H., Oliha, J. S., & Nwokediegwu, Z. Q. S. (2024). Review of smart water management: IoT and AI in water and wastewater treatment. *World Journal of Advanced Research and Reviews*, *21*(1), 1373–1382.

Deng, Y., & Zhao, R. (2015). Advanced oxidation processes (AOPs) in wastewater treatment. *Current Pollution Reports*, *1*, 167–176.

El-Gawad, H. A., Ebrahiem, E. E., Ghaly, M. Y., Afify, A. A., & Mohamed, R. (2023). An application of advanced oxidation process on industrial crude oily wastewater treatment. *Scientific Reports*, *13*(1), 3420.

El Sharkawy, H. M., Shawky, A. M., Elshypany, R., & Selim, H. (2023). Efficient photocatalytic degradation of organic pollutants over TiO2 nanoparticles modified with nitrogen and MoS2 under visible light irradiation. *Scientific Reports*, *13*(1), 8845.

Fane, A. G., Wang, R., & Jia, Y. (2010). Membrane technology: past, present and future. In *Membrane and desalination technologies* (pp. 1–45). Humana Press.

Fajobi, M. O., Lasode, O. A., Adeleke, A. A., Ikubanni, P. P., & Balogun, A. O. (2022). Effect of biomass co-digestion and application of artificial intelligence in biogas production: A review. *Energy Sources, Part A: Recovery, Utilization, and Environmental Effects*, *44*(2), 5314–5339.

Ghjair, A. Y., & Abbar, A. H. (2023). Applications of advanced oxidation processes (Electro-Fenton and sono-electro-Fenton) for COD removal from hospital wastewater: Optimization using response surface methodology. *Process Safety and Environmental Protection*, *169*, 481–492.

Giwa, A. S., Maurice, N. J., Luoyan, A., Liu, X., Yunlong, Y., & Hong, Z. (2023). Advances in sewage sludge application and treatment: Process integration of plasma pyrolysis and anaerobic digestion with the resource recovery. *Heliyon*, *9*(9), e19765.

Guven, H., Ersahin, M., Ozgun, H., Ozturk, I., & Koyuncu, I. (2023). Energy and material refineries of future: Wastewater treatment plants. *Journal of Environmental Management*, *329*, 117130.

Hicks, K. A., Fuzzen, M. L., McCann, E. K., Arlos, M. J., Bragg, L. M., Kleywegt, S., Tetreault, G. R., McMaster, M. E., & Servos, M. R. (2017). Reduction of intersex in a wild fish population in response to major municipal wastewater treatment plant upgrades. *Environmental Science & Technology*, *51*(3), 1811–1819.

Jaramillo, M. F., & Restrepo, I. (2017). Wastewater reuse in agriculture: A review about its limitations and benefits. *Sustainability*, *9*(10), 1734.

Kara, S., Hauschild, M., Sutherland, J., & McAloone, T. (2022). Closed-loop systems to circular economy: A pathway to environmental sustainability? *CIRP Annals*, *71*(2), 505–528.

Khalef, R. N., Hassan, A. I., & Saleh, H. M. (2022). Heavy metal's environmental impact. In H. M. Saleh & A. I. Hassan (Eds.), *Environmental Impact and Remediation of Heavy Metals*. IntechOpen, 2022, pp. 1–36. doi: 10.5772/intechopen.103907.

Li, X., Wei, H., Song, T., Lu, H., & Wang, X. (2023). A review of the photocatalytic degradation of organic pollutants in water by modified TiO2. *Water Science & Technology*, *88*(6), 1495–1507.

Madhav, S., Ahamad, A., Singh, A. K., Kushawaha, J., Chauhan, J. S., Sharma, S., & Singh, P. (2020). Water pollutants: Sources and impact on the environment and human health. In P. Singh, S. R. Mishra, & A. Kumar (Eds.), *Sensors in Water Pollutants Monitoring: Role of Material* (pp. 43–62). Springer. doi: https://doi.org/10.1007/978-981-15-0671-0_4.

Mafuta, C., Goncalves, A. S., Baker, E., Thygesen, K., Macmillan-Lawler, M., & Patterson, T. (2021). *Sanitation and Wastewater Atlas of Africa.* Abidjan, Nairobi, and Arendal: AfDB, UNEP, and GRID-Arendal.

Mahmood, S., Mahmood, N., Liaquat, M. U., Ikhlaq, F., Shehzad, H., & Yao, S. (2026). Electrodialysis for sustainable water desalination: Principles, applications, challenges, and future directions. *Journal of Environmental Chemical Engineering*, *14*, 121466.

Massoud, M. A., Tarhini, A., & Nasr, J. A. (2009). Decentralized approaches to wastewater treatment and management: Applicability in developing countries. *Journal of Environmental Management*, *90*(1), 652–659.

Mishra, S., Bharagava, R. N., More, N., Yadav, A., Zainith, S., Mani, S., & Chowdhary, P. (2019). Heavy metal contamination: An alarming threat to environment and human health. In *Environmental Biotechnology: For Sustainable Future* (pp. 103–125). Springer.

Mo, W., & Zhang, Q. (2013). Energy–nutrients–water nexus: Integrated resource recovery in municipal wastewater treatment plants. *Journal of environmental management*, *127*, 255–267.

Moslehyani, A., Mobaraki, M., Isloor, A. M., Ismail, A. F., & Othman, M. H. D. (2016). Photoreactor-ultrafiltration hybrid system for oily bilge water photooxidation and separation from oil tanker. *Reactive and Functional Polymers*, *101*, 28–38.

Nguyen, L. N., Kumar, J., Vu, M. T., Mohammed, J. A., Pathak, N., Commault, A. S., Sutherland, D., Zdarta, J., Tyagi, V. K., & Nghiem, L. D. (2021). Biomethane production from anaerobic co-digestion at wastewater treatment plants: A critical review on development and innovations in biogas upgrading techniques. *Science of the Total Environment*, *765*, 142753.

Obaideen, K., Shehata, N., Sayed, E. T., Abdelkareem, M. A., Mahmoud, M. S., & Olabi, A. G. (2022). The role of wastewater treatment in achieving sustainable development goals (SDGs) and sustainability guideline. *Energy Nexus*, 7, 100112. https://doi.org/10.1016/j.nexus.2022.100112.

Odirile, P. T., Marumoloa, P. M., Manali, A., & Gikas, P. (2021). Anaerobic digestion for biogas production from municipal sewage sludge: A comparative study between fine mesh sieved primary sludge and sedimented primary sludge. *Water*, *13*(24), 3532.

Parveen, F., & Khan, S. J. (2023). Wastewater treatment in Pakistan: Issues, challenges and solutions. In M. Ahmad (Ed.), *Water Policy in Pakistan: Issues and Options* (pp. 323–349). Springer.

Ponnusami, A. B., Sinha, S., Ashokan, H., Paul, M. V., Hariharan, S. P., Arun, J., Gopinath, K.P., Le, Q.H. & Pugazhendhi, A. (2023). Advanced oxidation process (AOP) combined biological process for wastewater treatment: A review on advancements, feasibility and practicability of combined techniques. *Environmental Research*, *237*, 116944.

Qasim, S. R., & Zhu, G. (2017). *Wastewater treatment and reuse, Theory and design examples, Volume 1: Principles and basic treatment* (1st ed.). CRC Press. https://doi.org/10.1201/b22368

Raouf, M. E. A., Maysour, N. E., Farag, R. K., & Abdul-Raheim, A. M. (2019). Wastewater treatment methodologies, review article. *International Journal of Environmental & Agricultural Science*, *3*, 018.

Rathi, B. S., Kumar, P. S., & Vo, D.-V. N. (2021). Critical review on hazardous pollutants in water environment: Occurrence, monitoring, fate, removal technologies and risk assessment. *Science of the Total Environment*, *797*, 149134.

Rodriguez, D. J., Serrano, H. A., Delgado, A., Nolasco, D., & Saltiel, G. (2020). From waste to resource. *Water Papers*. World Bank. https://doi.org/10.1596/33436

Rosemarin, A., Macura, B., Carolus, J., Barquet, K., Ek, F., Järnberg, L., Lorick, D., Johannesdottir, S., Pedersen, S. M., & Koskiaho, J. (2020). Circular nutrient solutions for agriculture and wastewater—A review of technologies and practices. *Current Opinion in Environmental Sustainability*, *45*, 78–91.

Roth, F. P., & Worthington, C. K. (2023). *Treatment Resource Manual for Speech-Language Pathology*. Plural Publishing.

Saravanan, A., Kumar, P. S., Jeevanantham, S., Karishma, S., Tajsabreen, B., Yaashikaa, P., & Reshma, B. (2021). Effective water/wastewater treatment methodologies for toxic pollutants removal: Processes and applications towards sustainable development. *Chemosphere*, *280*, 130595.

Sharma, A., Ahmad, J., & Flora, S. (2018). Application of advanced oxidation processes and toxicity assessment of transformation products. *Environmental Research*, *167*, 223–233.

Sniatala, B., Kurniawan, T. A., Sobotka, D., Makinia, J., & Othman, M. H. D. (2023). Macro-nutrients recovery from liquid waste as a sustainable resource for production of recovered mineral fertilizer: Uncovering alternative options to sustain global food security cost-effectively. *Science of the Total Environment*, *856*, 159283.

Tariq, A., & Mushtaq, A. (2023). Untreated wastewater reasons and causes: A review of most affected areas and cities. *International Journal of Chemical and Biochemical Sciences*, *23*, 121–143.

Vardhan, K. H., Kumar, P. S., & Panda, R. C. (2019). A review on heavy metal pollution, toxicity and remedial measures: Current trends and future perspectives. *Journal of Molecular Liquids*, *290*, 111197.

Villarín, M. C., & Merel, S. (2020). Paradigm shifts and current challenges in wastewater management. *Journal of Hazardous Materials*, *390*, 122139.

Wang, Q., Xia, C., Alagumalai, K., Le, T. T. N., Yuan, Y., Khademi, T., Berkani, M., & Lu, H. (2023). Biogas generation from biomass as a cleaner alternative towards a circular bioeconomy: Artificial intelligence, challenges, and future insights. *Fuel*, *333*, 126456.

Wid, N., & Raudin, N. (2023). Recovery of biogas from food waste using treated and untreated anaerobic digestion. *IOP Conference Series: Earth and Environmental Science*. (Vol. 1205, No. 1, p. 012004). IOP Publishing.

Xiao-jun, W., Jian-yun, Z., Elmahdi, A., Shahid, S., & Juan, G. (2023). A water resources assessment framework for management strategies of large coal-power bases development in China. *Mitigation and Adaptation Strategies for Global Change*, *28*(6), 35.

9 Transition to Water 5.0

9.1 INTRODUCTION TO WATER 5.0 AND THE AUTONOMOUS FUTURE WATER MANAGEMENT SYSTEMS

Industry 5.0, the next phase of the industrial revolution, is emerging from Industry 4.0, which combines industrial automation with cloud technology and data analytics to boost productivity. Unlike previous industrial revolutions, Industry 5.0 is not driven by breakthroughs or changing manufacturing practices. Instead, it advocates for a significant objective shift, prioritising societal value and long-term sustainability over revenue. This shift is crucial for a sustainable future for both businesses and society (Carayannis et al., 2023; Ali & Johl, 2024; Sharma & Gupta, 2024).

The European Commission claims that Industry 5.0 proposes an industrial paradigm beyond development and productivity, highlighting the industry's responsibilities and contribution to society and the environment (Pereira & dos Santos, 2023). As a result, the industry may play a key role in addressing our most pressing issues. Cloud-connected automation has altered manufacturing as part of Industry 4.0. It highlighted the value of data by improving process visibility, predictive analytics, and scenario planning. Emerging concepts like artificial intelligence (AI), machine learning, edge computing, the Internet of Things (IoT), and digital twins have entered the industrial lexicon and toolkit, resulting in operational efficiencies and improved insights for informed future decision-making when applied to production assets and processes. Industry 5.0 is primarily a call to action and pursuing a more significant purpose (Forward, 2024; Oeij et al., 2024; Satheesh & Shukla, 2024).

One of the first steps is understanding the industrial activity regarding natural resource use and potential environmental stresses (water, carbon, etc.) (Li & Li, 2024). This data foundation guides goal planning, making prioritising investments in cloud-connected automation or large-scale equipment infrastructure easier. The early implementation of sustainability-focused actionable dashboards at various organisational levels may help unify the organisation around shared sustainability goals. Water performance is critical in many industries and is our company's area of expertise; thus, we are constantly improving our ability to give Industry 5.0 solutions to assist customers in their aspirations. The 5.0 transition seeks to establish more organic connections between human and machine interactions in industrial environments. This initiates significant changes in the industrial sector, emphasising worker's health during manufacturing and employing innovative technology to promote prosperity beyond simple employment and economic growth while adhering to the planet's production limitations.

It builds on the Industry 4.0 paradigm by "explicitly positioning research and innovation to facilitate the transition towards a sustainable, human-centric, and resilient European industry" (Leng et al., 2022; Mourtzis et al., 2022; Golovianko et al., 2023; Martini et al., 2024). This transition is assisted by IoT and auxiliary technologies,

DOI: 10.1201/9781003633839-9

which allow for integrated cyber-physical convergence, digitisation, substantial automation, and information technology (IT) and operational technology (OT) operations synthesis. Supplementary tools and components include cyber-physical systems, horizontal and vertical integration, the transition from industrial automation to digital transformation, and networked value chains. Numerous benefits promised by Industry 4.0 advancements are only achievable at the advanced stages of adoption. The Industry 4.0 phased vision strongly emphasises autonomous and semi-autonomous manufacturing processes, hyper-automation, advanced robotics, self-optimising systems, and data exchanges; yet, these features may present complicated implementation challenges (Xu et al., 2018; Oztemel & Gursev, 2020). Ninety per cent of manufacturers prioritise the industrial use of digital tools; nevertheless, just one in four anticipates new revenue streams from present advancements. Thus, while many businesses have changed their business models to capitalise on the benefits of Industry 4.0, only a few have accomplished their full transformation goals. Although the most recent industrial innovation is based on 4.0 technology, Industry 5.0 aims to revitalise industrial digitisation by prioritising human operators, operational resilience, and true sustainability objectives. This new revolution redefines the advantages derived from technology, data, ecosystems, machinery, and the intelligent environments enabled by OT and IT. The fundamental reforms of Industry 5.0 extend beyond technology, focusing instead on enhancing industrial companies through three primary objectives:

- Human/Operator-centric (user experience design, operator experience, etc.)
- Resilience
- Sustainability

(ElFar et al., 2021; Mourtzis et al., 2022; Saniuk et al., 2022; Ghobakhloo et al., 2023)

9.1.1 Human/Operator-Centric (User Experience Design, Operator Experience, Etc.)

The cornerstone of Industry 5.0 success lies in technology and the individuals using it. Implementing new systems and solutions necessitates evaluating your team's operations and interactions with industrial and administrative processes to guarantee that the interfaces between humans and machines are more intuitive. Acquiring and maintaining talent has become a significantly greater challenge for several enterprises. The requisite skills have become prohibitive. Placing the user at the core of industrial technology adoption addresses labour shortages and optimises human potential (Maddikunta et al., 2022).

The success criteria of Industry 4.0 are predominantly centred on process efficiency and profit maximisation. The extensive supply chain disruption has underscored firms' need to prioritise operational resilience. This requires organisational flexibility and the ability to depend on versatile technologies. While agility and flexibility have emerged as organisational goals, these attributes do not inherently lead to enhanced resilience. As supply chain transparency and predictive forecasting demonstrate, digital technologies can anticipate crises and ensure consistent performance (Bibby & Dehe, 2018; Bai et al., 2020).

Genuinely sustainable enterprises focus on enhancing their positive influence rather than just mitigating their harmful effects. Ecological, social, and governance (ESG) corporate responsibility extends beyond mitigating adverse environmental impacts of company activities, urging manufacturers to contribute to the solution actively (Annesi et al., 2024; Kandpal et al., 2024; Mebar, 2024). Companies may respond to this transformation by implementing measures that enhance efficiency and minimise waste inherent in Industry 5.0 projects. As an organisation's digital systems grow increasingly intense and flexible, it can constantly select more sustainable solutions.

9.1.2 Introduction to Water 5.0

Water is the cornerstone of life and civilisation. Humanity's capacity to manage, distribute, and save water has changed throughout history, from ancient aqueducts to modern water treatment facilities. Over the ages, water management technology has advanced through numerous key milestones, each addressing the issues of its time. This chapter introduces the notion of Water 5.0, a visionary framework for the next stage of water technology evolution that incorporates cutting-edge solutions to solve the problematic issues of the 21st century. A cursory look at the history of water management reveals early civilisation and rudimentary supply systems throughout the First Industrial Revolution. Ancient civilisations such as Mesopotamia, Egypt, and the Indus Valley created primitive methods for diverting and distributing water. Canals, wells, and cisterns were significant inventions during the period. These systems were mainly manual, with little consideration of water quality and resource sustainability (Viollet, 2007; Angelakıs et al., 2020). The second revolution went a step further in response to the necessity for an industrial revolution and increased infrastructure. The Industrial Revolution saw advances in engineering, including the development of large-scale dams, pipelines, and urban water distribution systems (Angelakis et al., 2024). Water treatment technology, such as chlorination and filtration, appeared at this point to tackle public health concerns (Gitis & Hankins, 2018).

With the emergence of the Third Industrial (water) Revolution, there was a foundation for environmental awareness and regulatory frameworks. This occurred in the mid-20th century when water management became more mindful of water contamination and resource shortage. This period prioritised regulatory measures such as the Clean Water Act and technology solutions to decrease industrial and agricultural pollution. Digital transformation and smart water systems were invented during the Fourth Industrial Revolution. Sensors, IoT devices, and data analytics were brought into water management through digital technologies, whereas smart systems optimised resource allocation, enhanced leak detection, and enabled predictive maintenance for ageing infrastructure (Gohil et al., 2021; Sharma et al., 2024).

The need for efficient water use has reached new heights, whether driven by corporate social responsibility (CSR) duties or local water conservation rules. Water 5.0 symbolises a paradigm shift that builds on previous triumphs while addressing enormous concerns such as climate change, urbanisation, and global water inequity (Goede, 2020). It incorporates sophisticated technology, multidisciplinary approaches, and community-driven solutions. Furthermore, the significant association between industrial water use and energy consumption and the impact of rising costs and carbon emissions highlight the need to implement Industry 5.0 solutions for water management. A computerised water balance is one way to start the process. Using this technique, a firm may measure water usage over its whole footprint and study all water-related flows—inputs and outputs—at each facility. Implementing Industry 4.0 technologies such as networked sensors and instruments, digital dashboards, digital twin models, and predictive analytics can greatly enhance calculation precision and applicability. Maintaining a strong online water balance allows plant managers to identify opportunities for intervention to improve water efficiency or to design circular and closed-loop systems that recycle and reuse process water. Significant progress may be made when a facility begins autonomous data gathering and integrates this data with algorithms to improve each process, notably the operational efficiency of assets in water conservation (Arumugam et al., 2025).

The overarching plan starts at the macro level with the water balance and gradually narrows its focus through digital technologies. Digital technology enables informed decision-making. Furthermore, they automate some labour portions, allowing operators to solve future difficulties. Starting an Industry 5.0 journey does not necessarily need a company to invest heavily in new digital capabilities. Many people may already have specific data and the automated infrastructure required for advancement. Each company should analyse the need for new technology and calculate return on investment based on its unique circumstances. Companies are encouraged to apply a detailed technique to determine their water expenses. Organisations may improve their Industry 5.0 deployments by working with solution providers with domain expertise and leadership in digital technology. By combining existing data sources with Industry 4.0 technologies and analytics, businesses may identify optimum solutions and translate their data into actionable information, speeding their sustainability goals (Ren et al., 2019; Ikegwu et al., 2022).

9.1.3 Autonomous Future Water Management Systems

Water management has taken a new shape with the innovation of tools that make monitoring easy, especially with the ideals of Industry 4.0. Advanced technological water management uses information and communication technology to collect and analyse real-time water data. This technique is critical for addressing water-related concerns of various sizes and conditions. Integrating contemporary sensors and data analytics allows stakeholders to make informed decisions quickly, boosting water distribution efficiency and predictive maintenance of water infrastructure. These systems enable continuous monitoring and administration, ensuring water sustainability and resistance to unexpected environmental changes. An example of such devices is aerial drones, which have emerged as beneficial tools for monitoring sewage systems, offering a cost-effective and efficient method for inspecting inaccessible places. Fitted with high-resolution cameras and sensors, these drones can swiftly identify obstructions, leaks, and structural deficiencies in sewer systems, diminishing the necessity for manual inspections and lessening disruption. Drones permit real-time data collection, allowing utilities to address possible faults rapidly, boosting maintenance procedures, and increasing infrastructure longevity (Alsamhi et al., 2019; Spadoro et al., 2024).

Another tool is uncrewed aerial vehicles for aquatic quality assessment in reservoirs. Water quality in reservoirs is essential for guaranteeing safe drinking water and sustaining healthy ecosystems. Autonomous drones and advanced sensors improve water quality monitoring by offering continuous, real-time data on pH levels, turbidity, and temperature. These drones can independently roam reservoirs, spanning huge territories and accessing difficult or risky locations without human control. Another device invented for a similar purpose is the biomimetic drone for non-intrusive aquatic quality assessment. Drones, built to imitate fish or birds, are being constructed for non-invasive water quality evaluation, borrowing inspiration from nature. These biomimetic drones may integrate unobtrusively into natural settings, enabling data collection without harming species or ecosystems. These drones, mounted with sensors to analyse critical water quality indicators, offer a unique technique for monitoring, merging advanced technology with a decreased environmental effect.

9.2 POTENTIAL IMPACTS OF WATER 5.0 ON SOCIETY, ENVIRONMENT, AND ECONOMICS

Water 5.0 enhances the continuous provision of renewable energy. Renewable energy will become a substantial source of power generation shortly since these resources may be used repeatedly to provide usable energy. Energy resources are frequently classified as fossil resources, renewable resources, and nuclear energy resources. Renewable energy sources such as hydropower, wind, solar, biomass, ocean energy, biofuel, and geothermal account for 15–20% of worldwide energy supply. The globe is becoming a global village due to rising energy demands from a fast-growing population, necessitating the usage of fossil fuels such as coal, gas, and oil. This dependency creates unsustainable conditions and several concerns, such as fossil fuel depletion, environmental and geopolitical conflicts, the greenhouse effect, global warming, and fuel price volatility. Renewable energy is considered sustainable owing to its environmental friendliness and lower gas emissions; it also helps society economically, socially, and environmentally. "Approximately 1.6 billion individuals lack access to electricity, and around 1.1 billion are devoid of a water supply" (Behera et al., 2019; Pachauri et al., 2012). Renewable energy resources have the potential to meet global energy demand, protect the environment, and provide energy security. Despite the significant benefits of these resources, several problems remain, such as production unpredictability owing to seasonal swings, which is a major problem with wind and hydroelectric power facilities. To provide a lasting solution to this problem, Water 5.0 comes into play. With the aid of Water 5.0, wastewater can be treated and recycled to answer the seasonal swing problem of hydroelectricity supply (Kumar, 2020).

Water 5.0 also has an impact on the agricultural process. The agricultural industry is recognised as a substantial consumer of water resources, accounting for 70% of worldwide freshwater extraction to irrigate around 25% of arable land. Climate change and population growth make water supplies for agriculture more constrained. To address these challenges, Water 5.0, explained by water management methods and precision agriculture technologies, must be implemented to increase water consumption efficiency and satisfy agricultural production expectations despite a limited quantity of land and water resources (Martos et al., 2021; Zhang, 2024). Precision agriculture and intelligent irrigation technology enable farmers to improve resource efficiency while reducing plant water stress. Smart irrigation is the monitoring and management of water flow at ideal times, places, and amounts while considering soil moisture, climatic conditions, and plant health. Conventional irrigation practices may result in excessive or inadequate watering, leading to nutrient leaching, water waste, and declining quality and output. Improving irrigation efficiency is critical for reducing water use in agricultural activities while increasing crop yields and minimising environmental impact (Askaraliev et al., 2024). Improved irrigation may be achieved by introducing efficient irrigation methods such as drip or deficit irrigation, improving water application precision, changing agricultural practices such as crop rotation and conservation tillage, and correcting irrigation system leaks or damage. These technologies have the potential to significantly reduce the carbon footprint, reduce water use, decrease agricultural runoff, reduce energy needs for water pumping and transportation, and reduce irrigation-related extraction and transport costs. Drip irrigation is critical in sustainable agriculture because it accurately delivers water and nutrients to plant roots. Nonetheless, it encounters challenges, such as blockage and soil salinity. Subsurface irrigation efficiently solves these issues, reducing evaporation and weed growth while increasing safety through recovered wastewater. Desalination of wastewater also reduces salt buildup. Fertilisers are included in a drip irrigation system (known as fertigation), allowing for more precise nutrient delivery and resulting in higher crop yields and soil health (Shemer et al., 2023).

As an offshoot of Water 5.0, advanced hydro-economic decision support systems incorporate sophisticated hydrological simulation models, optimisation algorithms, uncertainty and risk assessments, AI, machine learning, real-time decision-making, statistical analysis, and socioeconomic, institutional, political, legal, ecological, and cultural factors. These systems guide the development of water infrastructure, and the formulation of public-private policies focused on the allocation and conservation of water quality across multiple scales within diverse research studies concerning water quality, hydropower generation, agricultural practices, damage mitigation from extreme hydroclimatic events (e.g., floods, droughts), water provision for industrial and residential use, water security, and water infrastructure resilience. The Drought Water Rights Allocation Tool is an example of a decision support system used in California's Russian River Watershed to maximise water allocation among riparian and appropriative water rights holders (Whittington, 2016; Laird, 2018). The AQUATOOL Decision Support System is used in Spain to plan and manage water resources in river basins, including water quality and environmental economics. Hydroelectric modelling refers to the economic advantage of water systems with multiple interdependent users, each with its own set of aims, restrictions, preferences, and risk profiles, achieved by systematic optimisation or modelling. Simulation models assess system responses to a variety of situations. Optimisation models, often known as mathematical programming, seek to determine optimal values for an objective function that describes a system by employing an appropriate optimisation technique or strategy. For example, issues can assess optimal water reservoir capacity and management (including water releases and allocation), administration of multiple-reservoir systems, the concurrent utilisation of water resources, water quality, hydropower production, agricultural yield, water security, the resilience of water infrastructure, and flood mitigation and practical restrictions (e.g., data availability or computational resources) (Ortiz-Partida et al., 2023).

9.3 DEEP LEARNING AND AUTONOMOUS DECISION-MAKING CAPABILITIES IN WATER SYSTEMS

Since the inception of the Harvard Water Programme in 1955, computer simulations have been essential in water management. Physically based models have been created to represent urban water systems (UWSs) with varying degrees of complexity. Nonetheless, these models have stalled due to the complexity of UWSs, difficulty defining modelling assumptions, data scarcity, significant processing power, and the human resources required for model development and maintenance. Machine learning, a subset of AI, is widely seen as a disruptive technology capable of revolutionising global economies, ecosystems, and societies. Machine learning will handle water-related challenges such as resource efficiency, water supply, pollution, floods, and drought, contributing to the United Nations' sustainable development goals. This technology will be critical in transforming the scientific field and operations of the water business. Deep learning, a subset of machine learning, represents a significant advancement in AI by utilising large, multi-layered artificial neural networks (ANNs) to evaluate massive raw data sets. Traditional machine learning systems, such as multi-layer perceptron neural networks, are limited in processing raw data and need domain expertise. Deep learning automates feature extraction across several levels of representation, translating raw data into more abstract forms. Deep learning approaches include Convolutional Neural Networks (CNNs), Long Short-Term Memory (LSTM), Autoencoders, Graph Neural Networks (GNNs), and Deep Reinforcement Learning (DRL). These algorithms have proved useful in fields such as image recognition and have been deployed in various industries, including healthcare and finance. The effectiveness of deep learning in the water business has been extensively recognised, as indicated by an increase in research publications, case studies, and applications (Khan et al., 2022; Liang et al., 2023).

Deep learning is widely used in urban water management for anomaly detection, system forecasting, asset appraisal, operational management, planning, and maintenance (Fu et al., 2022). Anomaly detection is a diagnostic analytics tool for detecting failure events like leaks, contamination, blockages, and cyber-attacks (Mohanty et al., 2024a, 2024b). System prediction and asset evaluation provide information on hydraulic or asset state. System operation, planning, and maintenance are optimisation concerns that determine the most effective solutions under certain constraints. Deep learning can address all essential difficulties in the life cycle of UWSs, yet few well-validated approaches or solutions are available. (Fu et al., 2022). Data and algorithmic development challenges hamper the evolution and use of deep learning approaches in the water business. CNNs, LSTM networks, and hybrid models are common solutions to these difficulties (Chen et al., 2023; Chen et al., 2024).

Diagnosing sewage faults has become popular due to the difficult and time-consuming nature of manually evaluating sewer surface conditions using CCTV video (Li et al., 2019; Rayhana et al., 2020). In recent years, CNNs have been widely used for automated tasks such as image classification, object identification, and semantic segmentation. Image classification involves training binary classification CNNs for specific fault types, whereas object detection identifies defects and their locations. Sewer obstruction levels are calculated using CNNs (Guo et al., 2018). Various approaches have been used to improve CNN efficiency and precision, including a cost-sensitive activation layer and cost-mean loss. Object identification models are used to determine the types and locations of faults in sewage inspection frames, as several problems may exist inside a single image. CNN-based techniques may be divided into two kinds: region-based or two-stage detection, which requires extracting areas of interest, and one-stage detection, which does not require such regions. Faster R-CNN is a two-stage detection technique used by Wang and Cheng (2018), whereas YOLOv3 is a one-stage network used for real-time automated sewage defect detection. YOLOv3 is best for on-site detection because of its fast speed, whereas Faster Region-based CNN (FR-CNN) is better for distant analysis due to its improved detection accuracy. Defect tracking is proposed based on the detection results of a quicker R-CNN, allowing the enumeration of errors over subsequent video frames (Liu et al., 2020; Fu et al., 2022).

Using CNN-based approaches, semantic segmentation models can locate and categorise defects and structural components in sewage pipelines. These algorithms have also been used to unroll CCTV footage, segment sewer defects, and determine the severity of operational and maintenance errors using semantic segmentation results. Recent research shows that Gated Recurrent Unit (GRU) and LSTM models outperform traditional ANNs in anticipating Combined Sewer Overflows (CSOs) water depth. Deep learning algorithms improve multi-site CSO prediction by combining geographical correlations from several places and temporal trends at individual sites. Sewer flow and water depth are frequently projected using rainfall data in conjunction with previous flow and water depth readings. LSTM networks can reduce the impact of missing observed data by including alternate input data, such as temporal parameters and precipitation intensity. Groundwater level data can improve the performance of LSTMs in estimating flow at several points within a sewage system. The higher predictive power of LSTMs in sewer flow modelling has been used to improve in-sewer storage management and operational resilience at wastewater treatment plants (WWTPs).

Recurrent Neural Networks (RNNs) and LSTM models have primarily been used to forecast critical state variables of urban wastewater systems, such as water quality, flow, and water level at various components and CSOs within the sewer system and the WWTP. Nonetheless, further research is needed to understand the complex connections between elements in wastewater systems and demand predictions. Deep learning is required to interpret high-resolution data in hydrodynamic flood modelling, including catchment and meteorological information. Nonetheless, the availability of such information is a considerable challenge in many places. Deep learning can manage large amounts of aerial images, Light Detection and Ranging (LiDAR), satellite, and radar weather data, resulting in multispectral data that improves urban flood prediction. CNNs, RNNs, autoencoders, and their hybrids are often used models for image segmentation, land classification, terrain attribute extraction, object identification, and multi-source image fusion. Deep learning algorithms have been used to assess disasters during floods using remote sensing photos. CNNs have been used to improve rainfall forecasting precision, estimate flood extent using data from uncrewed aerial vehicles, and monitor water depth using CCTV footage and crowd-sourced pictures. Precise high-resolution rainfall and flood depth representation provide improved data for calibrating urban flood models (Fu et al., 2022).

Because of the difficulty involved with evaluating large datasets of high-resolution urban catchment features and the scarcity of quantifiable flood and water infrastructure data, deep learning has received little attention in urban pluvial flood forecasting. Nonetheless, multiple studies suggest that CNNs may predict urban floods. Guo et al. (2021) developed a hybrid model that predicts the maximum flood depth for rainfall events using a convolutional autoencoder and a feedforward fully connected neural network coupled to the autoencoder's latent layer. Deep learning has been used more widely in data processing than flood prediction, with approaches based mostly on CNNs. Exploring alternate architectures, such as hybrid CNN-LSTM algorithms, is critical for improving flood-predicting accuracy in the future.

The US Department of Homeland Security has designated the water and wastewater industry as a primary target for cyberattacks among 16 critical infrastructure systems. Therefore, cybersecurity and asset monitoring are vital. In 2015, the United States reported 25 cybersecurity incidents, making it the third most targeted industry behind manufacturing and energy. Deep learning approaches, including LSTM, autoencoders, and GNN models, have significantly improved cyber-attack detection in high-dimensional datasets. An LSTM model was developed to detect cyberattacks in a water treatment system, and autoencoders were tested against 14 attack scenarios, including diverse components such as pumps, tanks, and controllers in a benchmark water distribution system (C-Town). The results showed that autoencoders outperformed standard machine learning approaches. A unique attention-based GNN approach outperformed baseline models, and a deep generative model for cyber-attacks was developed via variational inference after autonomously learning typical system behaviours from raw data. Prioritising research and innovation in UWS development is critical for mitigating the risks and vulnerabilities associated with water digitalisation and applying AI technologies in the water sector. Creating cybersecurity best-practice guidelines for UWSs is crucial since they are part of the critical national infrastructure.

Deep learning may be used as a soft sensor or surrogate model to monitor water assets (Fu et al., 2022; Zhang & Thorburn, 2022). 2D CNN and LSTM models were trained using one-minute operational data collected from 100 sensors over a year at a water treatment facility (Tipon Tanchangya et al. 2024). This results in much higher prediction accuracy than standalone CNN and LSTM models. GNNs have been used to replace missing pipe attribute data in wastewater systems, monitor fluctuations in the Fat-Oil-Grease layer, and predict pump sump problems. Cybersecurity and asset monitoring have been thoroughly investigated, and several strategies have been considered (Yaseen, 2024). DRL, an innovative real-time control and operational system, has been deployed in various areas, including water resource management and hydropower reservoir operations. Nonetheless, it has received little attention in UWSs due to its limited use in urban drainage systems and WWTPs (Yang et al., 2022).

DRL has been used in flood risk management by developing control systems for retention ponds. Mullapudi et al. (2020) developed a Deep Q-Network (DQN) based

DRL algorithm for real-time, non-predictive management of a distributed stormwater system with many detention ponds (Fu et al., 2022). Nonetheless, system-level management of many ponds was complex because of temporal dynamics, system interconnections, and high dimensionality. Policy-driven deep learning incorporating rainfall projections into decision-making has shown promise for improving flood management in urban drainage systems. Bowes et al. (2021) used a Deep Deterministic Policy Gradient (DDPG) actor-critic algorithm to regulate floods in urban drainage systems by manipulating valves at the base of retention ponds (Bowes et al., 2022). The method outperforms model predictive control and rule-based control systems by actively learning to adjust pond levels based on current and predicted circumstances. Reinforcement learning (RL) has also been used to reduce operating costs in WWTPs. Hernández-del-Olmo et al. (2016) demonstrated that RL can more effectively regulate the dissolved oxygen set points of a proportional-integral (PI) controller by taking into account system state variables, lowering operating costs when compared to manual operation and ammonium-based PI controllers. Based on Q-learning, RL was tested for controlling the advanced oxidation process of phenols using Fenton's reagent in a laboratory setting and enhancing the hydraulic retention length of anaerobic and aerobic reactors (Hernández-del-Olmo et al., 2018, 2019).

In addition to traditional RL, a policy-based DRL technique was created to minimise energy consumption by controlling pumps between primary and secondary treatment (Lee et al., 2022). A multi-agent DDPG approach was used to maintain dissolved oxygen and chemical doses in a wastewater treatment facility with continuous action and state spaces. In conclusion, numerous policy—and value-oriented DRL techniques have been applied in UWS system control; nevertheless, their applicability is limited, primarily to urban wastewater systems. Deep learning approaches for UWSs are still in their early stages of development, with most research focussing on benchmark networks, synthetic data, and laboratory or pilot systems. These works aim to demonstrate improved deep learning performance compared to traditional machine learning models or to promote a superior deep learning architecture by comparing alternative architectural designs. No applications have been deployed in the regular operation and management of real-world UWSs that exhibit substantial benefits or insights from the new use of deep learning technology (Chen et al., 2021; Joo et al., 2024; Negm et al., 2024).

Leakage identification is the most common challenge in deep learning research on urban water management, owing to data availability and the worldwide drive to reduce leakage and resource utilisation (Fu et al., 2022). Challenges have arisen due to insufficient massive data sets, labelling efforts, fault complexity, and challenges in providing direct support for pipe repair spending. Research should focus on developing deep learning approaches to improve sewer condition assessments that directly guide investment decisions. Contamination detection, albeit receiving substantial attention, is based on simplistic case studies and datasets, necessitating a breakthrough in resolving the complexities of actual pollution episodes before reaching the pinnacle of overblown expectations. The use of deep learning in asset monitoring has sparked great interest, considering recent advances in predictive maintenance. UWSs are naturally more complex, with several components and state variables driven by highly unpredictable environments and frequently have less data for deep learning training. DRL technology has proven improvements in system operation, although water research groups undervalue it. DRL has mostly been tested for flood control in sewage and wastewater treatment operations. Still, it has been used more widely in urban wastewater than in water distribution systems. Deep learning may outperform standard machine learning models in terms of efficiency and accuracy when used for classification, anomaly detection, and regression tasks in underwater sensor systems. Nonetheless, significant research gaps exist in extending deep learning approaches to improve process knowledge, increase the dependability, resilience, and sustainability of UWSs, and eventually build autonomous UWSs. Advances in research in these disciplines will help to close current gaps and improve the use of deep learning in industrial settings (Negm et al., 2024; Tipon Tanchangya et al., 2024).

9.4 KEY ENABLING TECHNOLOGIES: ADVANCED AI, ROBOTICS, AND AUTONOMOUS INFRASTRUCTURE

AI has a long history in urban water research, particularly planning, scheduling, optimisation, and exploration. Examples of water applications include operational scheduling and capacity increase optimisation (Krishnan et al., 2022; Fu et al., 2023). Machine learning, natural language processing, computer vision, and audio processing are some of the further AI areas. The taxonomy includes key AI subdomains and cross-cutting themes such as reasoning, planning, learning, communication, and perception. Reasoning, a hitherto under-represented topic in water applications, is gaining traction because of the necessity for explicit reasoning, causality, and explanation. The AI sub-domains are integrated; hence, their applications inside urban water infrastructure systems are interdependent. For example, the design of water distribution systems may use knowledge representation to include design concepts in the optimisation process, as well as machine learning as a surrogate model to improve optimisation efficiency. Common AI technologies in water management include machine learning algorithms that can assess prior water consumption data to estimate future demand, find irregularities, and upgrade water distribution systems, and neural networks that may depict complicated interactions in water systems, including the effect of climatic patterns on water supply. IoT devices and smart sensors acquire real-time data on water quality, flow rates, and environmental elements, which AI may evaluate to produce actionable insights. Lastly, it includes advanced data analytics techniques that allow the processing and understanding of enormous datasets, ultimately increasing decision-making and resource management (Paroha, 2022; Alshami et al., 2024; Mushtaque, 2024).

Pervasive Robotic Autonomous Systems will facilitate moving from reactive to proactive procedures, allowing for continuous and repetitive examination of pipe condition and operational performance. Robotic inspection and data collection are projected to supplement traditional data sources with significant extra information. As intelligence grows, so does the ability to learn, analyse, and anticipate network characteristics (Snelgrove et al., 2020; Mounce et al., 2021). Supplementary data presents fresh asset condition monitoring, performance evaluation, maintenance, and event analysis opportunities. Advanced robots increase operational efficiency and enable precise infrastructure management. Robotic technologies, such as autonomous robots and drones, inspect pipelines, detect leaks, and test water quality to eliminate manual interventions in hazardous or inaccessible environments. These robots navigate aquatic systems, doing real-time evaluations that spot potential issues before they escalate. Furthermore, robotic equipment integrated with AI and machine learning algorithms may predict maintenance needs and improve water distribution networks (WDNs) and treatment facility management (Ayari et al., 2016). Practical applications of advanced robotics in water management include inspection and maintenance. Autonomous robots go through water pipelines and infrastructure to assess structural integrity, locate leaks, and study degradation. These sensor-equipped robots provide complete, real-time data, reducing the need for human divers or manual inspections, especially in hazardous environments. Also, advanced robotics in water quality monitoring uses autonomous systems to collect and analyse water samples for contaminants, pH levels, and other quality indicators. These robots travel long distances to offer trustworthy and exact data instantly. Robotic systems enhance cleaning and treatment methods in water management. They purify reservoirs, tanks, and water channels while facilitating wastewater treatment. These robots painstakingly remove silt, garbage, and pollutants while working continuously in areas difficult for humans to reach (Walter et al., 2012; Bouch et al., 2022; Horne et al., 2023; Ranganathan et al., 2024).

Autonomous water infrastructure enables quick information retrieval from any point in the system, even inaccessible regions like pipes. The system monitors water and equipment, which improves operations and compliance while automating continuous controlled mitigation activities. Smart water infrastructure technologies include IoT devices (sensors), 5G connectivity, smart meters, real-time monitoring systems, advanced analytics, AI, machine learning algorithms, and data analytics (Rayhana et al., 2020; Fu et al., 2022). These devices and technologies work together to measure water consumption, predict infrastructure difficulties, and detect previously invisible contaminants by traditional monitoring methods. AI is a key aspect of intelligent water infrastructure. AI continuously promotes a better, more resilient future by reducing water loss, improving water quality and availability, and simplifying water distribution and treatment. Smart water infrastructure differs from traditional water management in two main ways: it collects far more data across the infrastructure. It is obtained in real or near-real time, allowing water management teams to respond quickly. Conventional water management systems collect data using various platforms, including supervisory control and data acquisition systems, computerised maintenance management systems, and laboratory information management systems (Therrien et al., 2020). Data fragmentation is a serious challenge, as it prevents timely comprehension of the entire picture. The broad deployment of 5G in various industrialised nations allows smart water infrastructure to collect more data since sensors operate in previously difficult-to-access sites. The collected data may be more comprehensive, and IoT devices can broadcast in near real-time across 5G-enabled WiFi networks. This faster access to data allows utilities to respond and avoid problems more quickly than before. For example, early leak detection reduces water loss, minimises service interruptions, and improves infrastructure resilience (Gohil et al., 2021; Dent, 2023; Thongtam et al., 2024).

9.5 AI-DRIVEN RESOURCE OPTIMISATION AND AUTONOMOUS LEAK DETECTION

Several sectors have found AI to be an important tool for streamlining procedures and operations. AI is a group of technologies that can process massive datasets independently, recognise patterns, and make data-driven decisions. These technologies include genetic algorithms, neural networks, and machine learning. AI-driven optimisation strategies offer the opportunity to boost efficiency, reduce costs, and reduce environmental impact by improving operational performance, reducing waste, and optimising resource utilisation (Bahramian et al., 2023). Rising populations, sophisticated economies, and shifting consumption habits are the key causes of the world's ever-increasing water demand (UNESCO, 2021). Water usage is connected to population expansion in several ways, including food production, electrical generation, industrial water use, and human consumption. According to the UNESCO (2019), water consumption is estimated to rise by 20–30% by 2050 due to projected global population growth to 8.5 billion in 2030 and 9.7 billion in 2050 (Boretti & Rosa, 2019; Islam, 2020). A 40% worldwide water shortage is anticipated by 2030 as freshwater sources dwindle and demand rises. Reduced demand in all water-related industries can help mitigate the consequences of this mismatch. Water utilities are immediately affected by technological or qualitative water losses, which significantly influence overall demand. Storage tank leaks, faulty pipes, and leaking service connections add to the losses. These elements cause various issues for . These include problems with service coverage assurances and intermittent circumstances, which can result in low pressures and pollutant penetration (du Plessis, 2017, 2023; Glied, 2020; Khilchevskyi & Karamushka, 2021).

Soil fluidisation produced by water leaks may cause damage to local infrastructure as well as increased expenditures for asset restoration, replacement, and system extension. Actual losses influence all operating expenses, including treatment supplies, energy, and exploitation (Serafeim et al., 2024). As mentioned earlier, the impacts may worsen if unprotected leaks are allowed to grow over time. As a result, water utilities are taking proactive steps to identify and resolve hidden vulnerabilities, and they are more aware of the need to respond swiftly and efficiently to reported breaches. The topic of leak detection and awareness has been extensively researched. There are two basic types of leak detection approaches: software-based and hardware-based. Methodologies based on hardware might be acoustic or non-acoustic. The former identifies leaks based on their acoustic qualities; however, it does not always function well with the increasingly ubiquitous plastic pipes, and it frequently necessitates substantial preparation and a large number of predetermined sensor placement sites (Spellman, 2013; Jouhara et al., 2017; Mudombi & Montmasson-Clair, 2020).

Leaks can restrict supply even if water distribution systems are intended to satisfy demand, owing to their interrelated components. To avoid financial losses, water companies must rapidly detect the presence of a leak and pinpoint its exact position (leak localisation). Conventionally, water supply can be improved by identifying leaks and repairing infrastructure. Several approaches for identifying and detecting leaks have been presented throughout the years (Zaman et al., 2020; Yuan et al., 2023). Modern

AI models for failure prediction and management planning might benefit from enormous amounts of historical data on water distribution systems. Predictive models can help predict equipment failure and detect important pipelines with leaks or flaws. Maintenance programmes may be improved to save time, money, and resources. Sarbu and Popa-Albu (2023) focused on optimal design, operation, and rehabilitation. Because ANNs do not function well with physical behaviour, another study on water pipe collapses suggested utilising support vector hybrid models (Vakilifard et al., 2018; Costache et al., 2024; Jayakumar et al., 2024).

9.6 CASE STUDIES ON EXPERIMENTAL WATER 5.0 APPLICATIONS

Brears and Brears (2020) assert that Singapore's Deep Tunnel Sewerage System (DTSS) functions as an underground channel for wastewater management. Consequently, routine inspections and maintenance are necessary to prevent operational delays. The tunnel's depth complicates manual inspections. Aero Lion Technologies intends to automate a miniature Uncrewed Aerial Vehicle (UAV) system for the purpose of capturing high-resolution images of the interior wall of the tunnel (Giron, 2019). The UAV will traverse the tunnel by employing maintenance holes and drop shafts to circumvent contact with flowing wastewater and its drag force. Aero Lion Technologies (ALT) has systematically developed an aerial platform to address these challenges. The patented UAV navigation system utilises depth cameras instead of range or image sensors in the absence of GPS. These cameras are meticulously engineered to capture both white colour and distance data for each pixel, as the structural uniformity of the DTSS interior inhibits range sensors from accurately estimating the UAV's forward velocity or position. In contrast, image sensors do not yield depth information from their two-dimensional images. The UAV's flight path within the DTSS will be fully automated, enabling it to conduct its inspection autonomously in the event of a communication failure with ground control. It will feature a ring of depth cameras and dynamic lighting to eradicate dead zones. Additionally, the UAV will employ sonar and/or laser rangefinders to assess the water and silt levels within the tunnel, compensating for its absence of vision-based navigation. After each expedition, the UAV's image data will be utilised to reconstruct an interior map of the tunnel and detect any issues. Drones provide a safe method for examining hazardous and unreachable areas of the sewage system (Brears, 2020; Anwar et al., 2022; Hyde-Smith et al., 2024).

9.7 SOLAR, WIND, AND OTHER RENEWABLE ENERGY SOURCES POWER WATER SYSTEMS

Approximately 78–80% of worldwide commercial energy comes from fossil fuels such as petroleum, coal, and natural gas. High-carbon sources negatively influence human health, land, air, precipitation, and the environment (Abas et al., 2015; Kalair et al., 2021). As a result, most governments have shifted their attention to low-carbon energy sources. In contrast to fossil fuels, depleted over time, renewable energy is a naturally abundant resource that may be used without jeopardising future energy requirements. Wind, solar, biomass, waves, and tidal energy are abundant renewable sources that may provide clean electricity. Recent advances in renewable energy technology have resulted in lower power-generating costs. Despite being recognised as an innovative power-generating approach, renewable energy is challenged by stochastic and unpredictable weather patterns. Its availability varies depending on the location. As a result, it is critical to supplement renewable energy sources with other sources such as batteries. Due to their intermittent nature, individual renewable energy sources are challenging to produce and operate (Khan et al., 2022). The primary goal is to increase electrical power output, reduce costs, offset the unfavourable adverse effects of fossil fuel burning, and improve overall system efficiency (Pranav et al., 2017).

Integrated renewable energy systems have recently gained popularity since a hybridised system may successfully serve end users with high-efficiency and consistent electricity instead of a single renewable source. A hybrid renewable energy system (HRES) can be used in standalone and grid-connected setups (Goel & Sharma, 2017). Autonomous systems require significant storage capacity to meet the demands. In grid-connected mode, storage capacity may be limited because insufficient power may be obtained from the grid. Power electronic controllers are needed in the grid-connected mode for load sharing, voltage regulation, harmonic management, and frequency control. The HRES working paradigm is divided into two categories: island mode, which uses generated power locally, and grid-connected mode, which connects the renewable energy source to the grid. Among renewable energy sources, wind power has had the fastest growth in global annual and cumulative installed capacity. Wind energy is standard worldwide, although its intensity varies by place. Unlike other renewable energy sources, wind energy may be used during the day or at night (Martínez Díaz, 2017; Basnet et al., 2023).

Solar energy is the cleanest and abundant renewable energy source. It is defined as the radiant light and heat emitted by the sun, which humans use through various technologies. The amount of energy collected from the sun depends on the radiation. Scientists classify radiation into two types: energy in wave form (electromagnetic waves) and energy in particle form (photons). The electromagnetic radiation emitted by the sun has a wavelength range of 0.1 nm to 104 m (Merry, 2001). Nonetheless, 95% of solar energy reaches Earth between the 0.3 and 2.4 μm band. Photons move at 3.0×10^8 m/s and have different energies measured in electron volts. Photovoltaic (PV) is derived from two terms: photo, meaning light, and voltaic, which refers to the unit of electric potential, the volt. PV cells, often known as solar cells, are semiconductors that convert sunlight into direct current electricity. A photovoltaic cell typically comprises a thin wafer with a surface layer of phosphorus-doped (N-type) silicon on top of a thicker layer of boron-doped (P-type) silicon. Doping creates the p-n junction and establishes an electric field on the cell's top surface. When photons from sunlight strike the photovoltaic cell, they cause electrons to migrate from the n-junction to the p-junction, generating an electric current. Regardless of size, a conventional silicon photovoltaic cell generates between 0.5 and 0.6 volts in an open circuit. The produced current is proportional to the amount of sunlight incident on the surface and the cell's efficiency and size (Wałachowska & Ignasiak-Szulc, 2021).

The solar cell is frequently connected in a series or parallel circuit, as in Figure 9.1, to create the required quantity of electricity. Solar modules are made up of solar cells

FIGURE 9.1 Battery bank installation showing multiple deep-cycle storage units connected in a parallel configuration for backup power supply. The figure illustrates the physical layout, terminal connections, and safety considerations associated with on-site energy storage systems.

encased in a protective laminate that is meant to withstand environmental conditions. The solar module is a critical component of any photovoltaic system. Solar panels are made up of one or more solar modules that have been pre-wired and may be installed in the field or on rooftops, as in Figures 9.2 and 9.3. A photovoltaic array is a pre-installed system for generating electricity that consists of a group of solar panels. The efficacy of photovoltaic modules and arrays is typically tested based on their maximum power output (W) under standard test conditions (Amran et al., 2024).

A new wind, water, and solar (WWS) energy infrastructure must provide on-demand energy as reliably as the current system (Delucchi & Jacobson, 2011; Jacobson et al., 2018). Any energy system must adjust to demand fluctuations across seconds, minutes, hours, seasons, years, and unexpected variations in producing supplies. Currently, electricity-system operators use "automatic generation control" (AGC) for frequency regulation to address fluctuations within seconds to a few minutes, spinning reserves for variations lasting minutes to an hour, and peak-power generation for hourly volatility (Beauvais et al., 2012; Harpool, 2018).

AGC and spinning reserves incur little cost, often accounting for less than 10% of total power costs. Despite increased wind power generation, costs will likely remain low; nevertheless, peak-power production may incur significant expenditures. The key difficulty for the existing energy system is the variability of electric power demand throughout the day and year, while the bulk of supply sources (coal, nuclear, and geothermal) stay stable. Peak- and gap-filling resources like natural gas and hydropower must be used to close the gap (Emmanuel et al., 2020).

A significant challenge to the present system is that severe incidents and unplanned maintenance may result in unexpected plant shutdowns. Unplanned maintenance may render coal plants inoperable, extreme heat waves can raise cooling water temperatures to the point where nuclear reactors must shut down, supply disruptions can

FIGURE 9.2 Close-up view of rooftop solar modules connected in parallel, increasing array current while maintaining a constant system voltage.

FIGURE 9.3 Solar panels installed on the rooftop of a residential building, illustrating a typical small-scale rooftop solar photovoltaic system.

limit natural gas availability, and droughts can reduce hydroelectric power output (Bach et al., 2013; Hansen, 2024). A WWS electrical system brings new challenges and opportunities for regularly meeting energy demands. WWS technology frequently has less downtime than modern electric power options. Between 2000 and 2004, the average coal plant in the United States saw 6.5% annual downtime due to unscheduled maintenance and 6.0% for scheduled maintenance (Bhutto et al., 2013; Gulagi, 2023). In contrast, modern wind turbines have downtime of roughly 0–2% on land and 0–5% offshore (Cevasco et al., 2021; Taboada et al., 2021). Commercial solar projects are expected to have an average downtime of 1%, with some reporting no downtime in a year and others seeing up to 10%. Furthermore, a significant difference exists between outages of centralised power plants (coal, nuclear, and natural gas) and distributed plants (wind, solar, and wave). Only a small portion of the electrical output is affected when individual solar panels or wind turbines fail. In contrast, the failure of a centralised plant affects a large portion of the network (Ramirez-Meyers et al., 2021; Smith et al., 2022).

Furthermore, the entire national grid may be affected if numerous important, centralised facilities go offline simultaneously due to a common problem. According to the Nuclear Power Daily, on November 2, 2009, one-third of France's nuclear power plants were offline due to a maintenance and refuelling backlog, prompting the country's power distribution firm to indicate that it may need to import energy from neighbouring markets for two months beginning in mid-November (Tai et al., 2022; Teisho, 2022). The primary issue is that the maximum solar or wind power at a given location varies over minutes, hours, and days, and this fluctuation frequently does not correspond to the demand pattern over the same time intervals. Other WWS technologies are stable over these time scales: tidal power is dependable due to tide predictability; geothermal energy supply is typically consistent; and hydroelectric power can be quickly activated or deactivated, currently serving to provide peaking and gap-filling power, though hydropower availability varies seasonally and annually. As a result, there will be times when a single installation cannot create enough power to meet demand and when it generates surplus power beyond needs, resulting in an economic inefficiency in producing capacity. Nonetheless, Delucchi and Jacobson (2011) proposed at least seven approaches for developing and managing a WWS energy system to guarantee that demand is regularly satisfied while avoiding excessive underutilised capacity:

a. link geographically scattered, naturally variable energy sources (e.g., wind, solar, wave, and tidal);
b. use a non-variable energy source, such as hydroelectric power, to bridge temporary gaps between demand and wind or solar supply;
c. implement "smart" demand-response management to modify flexible loads to correspond with the availability of WWS electricity;
d. store electric power at the producing site for later use;
e. increase WWS peak generation capacity to reduce instances where WWS power falls short of demand and allocate surplus power for hydrogen production for versatile transportation and heating applications;
f. use electric vehicle batteries for energy storage; and
g. predict weather patterns to improve energy supply planning.

9.8 ENERGY-EFFICIENT DESALINATION AND WATER TREATMENT

Liang et al.'s (2020) literary work concluded that water scarcity has developed as a barrier to social progress and economic prosperity, worsened by human activity and climate change. The global water demand is expected to surpass sustainable freshwater supply by 40% by 2030. As a result, the International Desalination Association notes that seawater desalination technologies, which extract freshwater from the vast ocean, have undergone substantial investigation. The desalination market has expanded significantly, with the total worldwide installed desalination capacity reaching 97.4 million m^3/d and the global contracted capacity reaching 104.7 million m^3/d in 2018 (Ahmed et al., 2021; Lin et al., 2021; Zapata-Sierra et al., 2021; Ghazi et al., 2022). Reverse osmosis (RO) currently accounts for around 65% of total desalinated water due to its inherent advantages, such as low cost and energy consumption (Feria-Díaz et al., 2021; Lim et al., 2021; Matin et al., 2021).

Desalination removes salt and other impurities from seawater, producing drinkable water, which is becoming increasingly important due to global water scarcity (Ayaz et al., 2022; Khondoker et al., 2023). To accomplish this, various methods have been developed, the most common of which is RO. This desalination technology uses a semi-permeable membrane to separate salt and pollutants from water. Recent research has focused on increasing the efficiency and sustainability of desalination processes, particularly by using renewable energy sources such as solar and wind power (Shokri & Fard, 2022; Alghassab, 2024). One important research area is the combination of solar energy with RO systems. According to research, using solar energy can significantly reduce energy expenditures associated with desalination procedures. A review found many preheating strategies that can improve the efficiency of solar-powered RO systems by raising the feedwater temperature, hence increasing production and cutting energy costs (Al-Obaidi et al., 2022; Shalaby et al., 2022; Alawad et al., 2023). Furthermore, concerns about brine disposal, a byproduct of desalination, have been addressed, emphasising the importance of cost-effective disposal methods, particularly for inland desalination operations. An alternative approach combines RO with multi-effect distillation (MED) systems, employing solar combined heat and power systems. MED is a thermal desalination technology that uses numerous rounds of evaporation and condensation to produce fresh water (Shalaby et al., 2022; Imandoust et al., 2024).

A techno-economic study of hybrid systems indicated that combining RO with MED can significantly lower the cost of freshwater production, making it a more viable option for needy populations (Ali et al., 2021; Al-hotmani et al., 2022). This integration maximises resource utilisation and extends operational hours through thermal energy storage. This technology stores thermal energy for later use, assisting in the balance of energy supply and demand while enhancing overall system efficiency (Vahidhosseini et al., 2024). Several renewable energy sources, including wind energy, have been examined for their potential application in desalination processes. Wind energy systems can provide a clean and efficient power source for RO and electrodialysis desalination, particularly in arid regions where water scarcity is severe. Combining wind turbines and desalination technology improves dependability and reduces the intermittent nature of wind energy, making these systems more robust and economically feasible (Greco et al., 2021). A complete evaluation of desalination technologies in conjunction with renewable energy sources has been offered, highlighting the need to select the optimal technological combinations to meet water demand sustainably. Studies found that the most efficient systems usually combine renewable energy sources and may use energy storage techniques to provide a consistent energy supply (Kebede et al., 2022). Recent studies focused on the environmental impact of desalination techniques. As the demand for freshwater rises, the need for sustainable and energy-efficient desalination technologies grows (Ahmed et al., 2021; Ihsanullah et al., 2021; Ayaz et al., 2022). Research has identified many methods and technologies for reducing the environmental effects of desalination, ensuring that it serves as a long-term solution to water scarcity rather than a temporary fix (Ayaz et al., 2022; Yatawatta & Sridarran, 2024).

9.9 THE ROLE OF WATER 5.0 IN SUSTAINABLE, CLIMATE-RESILIENT WATER SYSTEMS

Industry 5.0 should be viewed as a development. Furthermore, different organisations are at various stages of this process. Some have an edge because of a better awareness of the growing importance of environmental stewardship and the use of ESG reporting metrics (Leng et al., 2022; Dwivedi et al., 2023). Many organisations have created CSR projects, which may include water conservation efforts or commitments to use renewable energy. Industry 5.0 provides a catalyst and framework for improving their sustainability performance. For new businesses, Industry 5.0 provides a framework for developing sustainability-focused projects or resource-efficient procedures. One of the first steps they may take is understanding how they use natural resources and environmental pressures (Atiku et al., 2024). The carbon footprint is a commonly used metric. Companies may also attempt to quantify their greenhouse gas emissions or the rubbish produced. Another goal is to establish a water balance and precisely calculate water costs. Organising these data lays the groundwork for creating quantifiable sustainability performance goals, which may be strengthened with the support of the Industry 4.0 technical toolbox. Over 110 countries have pledged to become carbon neutral by 2050 in response to the urgent call for action on climate change. Several countries are strengthening their national climate targets in preparation for the United Nations Climate Change Conference (COP26). Climate change involves rapid collective action and careful assessment of how water may help address the interwoven concerns of climate and ecology while achieving the Sustainable Development Goals (De La Peña et al., 2022; Pawłowski & Rydzewski, 2024).

Jain et al. (2024) claimed climate change has a global influence on water resources, jeopardising basic needs and human rights to water and sanitation. Four billion people face severe water shortages for at least one month each year. Disasters typically appear as water-related disasters. Floods, landslides, storms, and droughts are becoming more frequent and severe, endangering human lives, critical ecosystems, and our economic and social development efforts (Shahid & Ahmad, 2023). Two hurricanes, Eta and Iota, have battered Central American regions in the last two weeks, exacerbating the already dire humanitarian situation and increasing the vulnerability of indigenous communities who lack access to clean water and sanitation (Ayaz et al., 2022; Jain et al., 2024). According to Morgan Stanley, global climate change has cost the economy around US$650 billion over the last three years. The COVID-19 outbreak has exposed significant socioeconomic disparities and vulnerabilities, most notably in access to water. Although handwashing is an important step in avoiding disease transmission, 3 billion people do not have access to handwashing facilities at home, and one of every four healthcare facilities lacks essential water services (World Health Organization, 2021).

Most governments understand the risks and the need to improve adaptability; of the 137 countries that included adaptation in their 2015 national climate pledges under the Paris Agreement Nationally Determined Contribution (NDCs), water emerged as the top priority for adaptation activities (Attoh et al., 2024; Sarpong, n.d.). Through the Climate Promise, United Nations Development Programme (UNDP) is aiding 115 countries in strengthening their climate pledges. They aid several nations that have decided to strengthen water-related measures to reduce emissions or adapt to their impact. Furthermore, some of UNDP's most important water-related activities are sponsored via the Climate Change Adaptation portfolio (Mehrotra & Benjamin, 2022; Pill, 2022; Qi & Qian, 2023). However, one aspect that is usually overlooked is water's contribution to the emission reduction initiatives governments promote. Nature-based solutions, such as green infrastructures, may improve carbon sequestration and replenish water sources. Nonetheless, climate change can harm water resources (Bayulken et al., 2021; Cassin & Matthews, 2021; Gooden & Pritzlaff, 2021).

Transitions in energy composition towards hydropower or biofuels may face restrictions owing to water availability and competing resource demands. The UNDP and its partners assist countries in assessing the interrelationships between water's many functions while strengthening their NDCs. Water and wastewater utilities have been claimed to account for 3–7% of national greenhouse gas emissions, both directly and indirectly (Du et al., 2023; Song et al., 2024). Direct emissions include methane and nitrous oxide produced during wastewater treatment and, probably more importantly, improperly treated waste and wastewater. Indirect emissions are created by the energy used in water extraction, transit, and treatment. This is driven by our economy's high demand for water and considerable inefficiencies in water supply and distribution. According to the World Bank, over 45 million cubic meters of water are squandered daily, resulting in an annual economic loss of more than $3 billion in developing countries. Rectifying these inefficiencies will boost millions of people's access to water, save water and energy, and reduce emissions from water and wastewater treatment operations (Mishra, 2023; Emran et al., 2024).

The current economic, environmental, and public health challenges provide a chance to improve water resource management and increase access to water and sanitation services. Collaboration among governments, civil society, and the private sector may expand services to the 785 million people who did not have access to essential water in 2017 and the 2 billion who still do not have proper sanitation. As countries establish recovery strategies and prepare to encourage behavioural changes for improved hygiene through investments in knowledge, accessibility, and infrastructure, we advocate for these investments to ensure a sustainable, resilient, and equitable development trajectory. The UNDP, in collaboration with the Green Climate Fund and the Global Environment Facility—Least Developed Countries Fund, is promoting climate change adaptation initiatives such as rainwater harvesting and service delivery to remote atolls in Small Island Developing States such as the Solomon Islands, Maldives, and the Marshall Islands (Garschagen & Doshi, 2022; Omukuti et al., 2022). Thanks to the United Nations Development Programme Global Environment Fund (UNDP-GEF) International Waters relationship, we are a major worldwide player in integrated water resource management within a transboundary framework. These programmes promote climate change adaptation and mitigation by employing an ecosystem-based approach. Water practitioners adapt to and mitigate climate change through the International Capacity Development Network for Sustainable Water Management, with solar-powered water systems being one of the most popular training programmes (Nandineni & Dash, 2022; Nkhata,

2022; Scheihing et al., 2022). To assist nations in effectively integrating water into the enhancement of their national climate commitments, UNDP and its collaborators (including Alliance for Global Water Adaptation (AGWA), Global Water Partnership (GWP), and Stockholm International Water Institute (SIWI) through the UNDP-SIWI Water Governance Facility partnership) have held a series of water and climate workshops to improve coordination among climate and water policymakers and experts (Yasin et al., 2021; Swedenborg et al., 2022). Supplementary workshops in a variety of languages and locations are now being prepared. UNDP operates in 170 countries and territories and will continue collaborating to combine resources and knowledge to improve access to essential services and promote equitable, sustainable resource management (Yu et al., 2024).

9.10 CHAPTER SUMMARY

This chapter looks at the transition to Water 5.0, an innovative water management paradigm that uses advanced technologies like IoT, AI, robotics, and renewable energy to address global water concerns sustainably. It highlights the development of water management from ancient aqueducts to the digital revolution of Industry 4.0, which culminates in the human-centric, resilient, and sustainable paradigm of Industry 5.0. This chapter emphasises the importance of autonomous systems in improving water use, strengthening infrastructure resilience, and ensuring resource efficiency through smart irrigation, precision agriculture, and renewable energy-driven desalination technologies. Case studies demonstrate the practical use of drones for sewage surveillance and HRESs for water desalination. This chapter investigates the ability of deep learning and AI to optimise UWSs, detect leaks, anticipate floods, and improve wastewater treatment, hence promoting sustainable and climate-resilient water infrastructure. The transition to Water 5.0 represents a significant shift in water resource management, ensuring long-term sustainability and resilience to climate change, population growth, and urbanisation. Water 5.0 uses AI, robotics, and the IoT to provide actionable insights, increase operational efficiency, and promote the development of climate-resilient and sustainable water infrastructure. Water 5.0 addresses important global concerns with advancements such as intelligent irrigation, renewable energy-powered desalination, and deep learning applications, paving the way for a future of more intelligent, efficient, and environmentally sustainable water management. This shift necessitates cross-sector collaboration to connect technological advancements with long-term development goals, ensuring fair access to water resources for all.

REFERENCES

Abas, N., Kalair, A., & Khan, N. (2015). Review of fossil fuels and future energy technologies. *Futures*, *69*, 31–49.

Ahmed, F. E., Khalil, A., & Hilal, N. (2021). Emerging desalination technologies: Current status, challenges and future trends. *Desalination*, *517*, 115183.

Al-hotmani, O., Al-Obaidi, M. A., John, Y., Patel, R., & Mujtaba, I. M. (2022). A hybrid multi effect distillation and double reverse osmosis system for most economical brackish water desalination. In L. Montastruc, & S. Négny (Eds.), *Computer Aided Chemical Engineering* (Vol. 51, pp. 133–138). Elsevier.

Al-Obaidi, M. A., Zubo, R. H., Rashid, F. L., Dakkama, H. J., Abd-Alhameed, R., & Mujtaba, I. M. (2022). Evaluation of solar energy powered seawater desalination processes: A review. *Energies*, *15*(18), 6562.

Alawad, S. M., Mansour, R. B., Al-Sulaiman, F. A., & Rehman, S. (2023). Renewable energy systems for water desalination applications: A comprehensive review. *Energy Conversion and Management*, *286*, 117035.

Alghassab, M. A. (2024). A review of hybrid solar desalination systems: Structure and performance. *Water Science & Technology*, *89*(5), 1357–1381.

Ali, E., Orfi, J., AlAnsary, H., Soukane, S., Elcik, H., Alpatova, A., & Ghaffour, N. (2021). Cost analysis of multiple effect evaporation and membrane distillation hybrid desalination system. *Desalination*, *517*, 115258.

Ali, K., & Johl, S. K. (2024). Driving sustainability in industry 5.0 through sociotechnical approach of quality management. *Total Quality Management & Business Excellence*, *35*(13–14), 1567–1592.

Alsamhi, S. H., Ma, O., Ansari, M. S., & Almalki, F. A. (2019). Survey on collaborative smart drones and internet of things for improving smartness of smart cities. *IEEE Access*, *7*, 128125–128152.

Alshami, A., Ali, E., Elsayed, M., Eltoukhy, A. E., & Zayed, T. (2024). IoT innovations in sustainable water and wastewater management and water quality monitoring: A comprehensive review of advancements, implications, and future directions. *IEEE Access*, *12*, 58427–58453. https://doi.org/10.1109/ACCESS.2024.3392573

Alshami, A., Ali, E., Elsayed, M., Eltoukhy, A. E., & Zayed, T. (2024). IoT innovations in sustainable water and wastewater management and water quality monitoring: A comprehensive review of advancements, implications, and future directions. *IEEE Access*, *12*, 58427–58453. https://doi.org/10.1109/ACCESS.2024.3392573

Amran, N. F. Q., Zulkifli, W. N. A. A. W., Jaalam, N., Arith, F., & Shah, A. S. M. (2024). Investigation of standard test condition requirement in establishing alternative measurement platform for photovoltaic cell. *Journal of Physics: Conference Series*, *2928*, 012003. https://doi.org/10.1088/1742-6596/2928/1/012003

Angelakis, A. N., Baba, A., Valipour, M., Dietrich, J., Fallah-Mehdipour, E., Krasilnikoff, J., Bilgic, E., Passchier, C., Tzanakakis, V. A., & Kumar, R. (2024). Water Dams: From ancient to present times and into the future. *Water*, *16*(13), 1889.

Angelakıs, A. N., Zaccaria, D., Krasilnikoff, J., Salgot, M., Bazza, M., Roccaro, P., Jimenez, B., Kumar, A., Yinghua, W., & Baba, A. (2020). Irrigation of world agricultural lands: Evolution through the millennia. *Water*, *12*(5), 1285.

Annesi, N., Battaglia, M., Ceglia, I., & Mercuri, F. (2024). Navigating paradoxes: Building a sustainable strategy for an integrated ESG corporate governance. *Management Decision*, *63*(2), 531–559. https://doi.org/10.1108/MD-10-2023-2006

Anwar, A., Liu, X., & Zhang, L. (2022). Biogenic corrosion of cementitious composite in wastewater sewerage system—A review. *Process Safety and Environmental Protection*, *165*, 545–585.

Arumugam, S. K., Tyagi, A. K., Tiwari, S., & Kumari, S. (2025). The role of futuristic technologies in building sustainable environment. In B. Sundaramurthy, & P. Dayalan (Eds.), *Creating AI Synergy through Business Technology Transformation* (pp. 247–276). IGI Global.

Askaraliev, B., Musabaeva, K., Koshmatov, B., Omurzakov, K., & Dzhakshylykova, Z. (2024). Development of modern irrigation systems for improving efficiency, reducing water consumption and increasing yields. *Machinery & Energetics*, *15*(3), 47–59. https://doi.org/10.31548/machinery/3.2024.47

Atiku, S. O., Jeremiah, A., Semente, E., & Boateng, F. (2024). *Eco-Innovation and Sustainable Development in Industry 5.0*. IGI Global.

Attoh, E. M., Amarnath, G., Sahana, V., Okem, A., Panjwani, S., Shashanka, G., Ukwattage, S., Seelanatha, K., & Alahacoon, N. (2024). Development of climate smart governance dashboard to support national climate action and resilience building. International Water Management Institute (IWMI); CGIAR Initiative on Climate Resilience. https://cgspace.cgiar.org/items/5089e162-7953-4622-b0ec-0675da172d47

Ayari, N., Chibani, A., Amirat, Y., & Matson, E. (2016). A semantic approach for enhancing assistive services in ubiquitous robotics. *Robotics and Autonomous Systems*, *75*, 17–27.

Ayaz, M., Namazi, M., ud Din, M. A., Ershath, M. M., & Mansour, A. (2022). Sustainable seawater desalination: Current status, environmental implications and future expectations. *Desalination*, *540*, 116022.

Bach, C., Gupta, A. K., Nair, S. S., & Birkmann, J. (2013). *Critical infrastructures and disaster risk reduction.* National Institute of Disaster Management and Deutsche Gesellschaft für internationale Zusammenarbeit GmbH (GIZ), New Delhi (p. 72). ISBN 978-3-944152-13-4.

Bahramian, M., Dereli, R. K., Zhao, W., Giberti, M., & Casey, E. (2023). Data to intelligence: The role of data-driven models in wastewater treatment. *Expert Systems with Applications*, *217*, 119453.

Bai, C., Dallasega, P., Orzes, G., & Sarkis, J. (2020). Industry 4.0 technologies assessment: A sustainability perspective. *International Journal of Production Economics*, *229*, 107776.

Basnet, S., Deschinkel, K., Le Moyne, L., & Péra, M. C. (2023). A review on recent standalone and grid integrated hybrid renewable energy systems: System optimization and energy management strategies. *Renewable Energy Focus*, *46*, 103–125.

Bayulken, B., Huisingh, D., & Fisher, P. M. (2021). How are nature based solutions helping in the greening of cities in the context of crises such as climate change and pandemics? A comprehensive review. *Journal of Cleaner Production*, *288*, 125569.

Beauvais, D., Prieur, A., & Bouffard, F. (2012). Smart Grid for balancing renewable energies–contributing distributed energy resources (Report No. 2012-177, RP-TEC 411-FLEXIN). CanmetENERGY, Natural Resources Canada.

Behera, B. K., Rout, P. K., Behera, S., Behera, B. K., Rout, P. K., & Behera, S. (2019). Water, energy and food security: Pillars for zero Hunger. In B. K. Behera, P. K. Rout, & S. Behera (Eds.), *Move towards Zero Hunger* (pp. 37–60). Springer. https://doi.org/10.1007/978-981-32-9800-2_2

Bhutto, A. W., Bazmi, A. A., & Zahedi, G. (2013). Greener energy: Issues and challenges for Pakistan—wind power prospective. *Renewable and Sustainable Energy Reviews*, *20*, 519–538.

Bibby, L., & Dehe, B. (2018). Defining and assessing industry 4.0 maturity levels–case of the defence sector. *Production Planning & Control*, *29*(12), 1030–1043.

Boretti, A., & Rosa, L. (2019). Reassessing the projections of the world water development report. *NPJ Clean Water*, *2*(1), 15.

Bouch, C. J., Shrimpton, E. A., Makana, L., Bowman, B., & Rogers, C. D. (2022). Robotic autonomous asset management: Benefit/value-based business model creation. *Infrastructure Asset Management*, *10*(2), 92–102.

Bowes, B. D., Wang, C., Ercan, M. B., Culver, T. B., Beling, P. A., & Goodall, J. L. (2022). Reinforcement learning-based real-time control of coastal urban stormwater systems to mitigate flooding and improve water quality. *Environmental Science: Water Research & Technology*, *8*(10), 2065–2086.

Bowes, B. D., Tavakoli, A., Wang, C., Heydarian, A., Behl, M., Beling, P. A., & Goodall, J. L. (2021). Flood mitigation in coastal urban catchments using real-time stormwater infrastructure control and reinforcement learning. *Journal of Hydroinformatics*, *23*(3), 529–547. https://doi.org/10.2166/hydro.2020.080

Brears, R. C. (2020). *Developing the Circular Water Economy*. Springer.

Brears, R. C. (2020). Singapore's public utilities board developing the circular water economy. In R. C. Brears, *Developing the circular water economy* (pp. 171–183). Palgrave Macmillan. https://doi.org/10.1007/978-3-030-32575-6

Carayannis, E. G., Canestrino, R., & Magliocca, P. (2023). From the dark side of industry 4.0 to society 5.0: looking "beyond the box" to developing human-centric innovation ecosystems. *IEEE Transactions on Engineering Management*, *71*, 6695–6711.

Cassin, J., & Matthews, J. H. (2021). Nature-based solutions, water security and climate change: Issues and opportunities. In J. Cassin, J. H. Matthews, & E. López Gunn (Eds.), *Nature-Based Solutions and Water Security* (pp. 63–79). Elsevier.

Cevasco, D., Koukoura, S., & Kolios, A. (2021). Reliability, availability, maintainability data review for the identification of trends in offshore wind energy applications. *Renewable and Sustainable Energy Reviews*, *136*, 110414.

Chen, K., Wang, H., Valverde-Pérez, B., Zhai, S., Vezzaro, L., & Wang, A. (2021). Optimal control towards sustainable wastewater treatment plants based on multi-agent reinforcement learning. *Chemosphere*, *279*, 130498.

Chen, M., Qian, Z., Boers, N., Jakeman, A. J., Kettner, A. J., Brandt, M., Kwan, M.-P., Batty, M., Li, W., & Zhu, R. (2023). Iterative integration of deep learning in hybrid Earth surface system modelling. *Nature Reviews Earth & Environment*, *4*(8), 568–581.

Chen, Q., Yang, Y., Qi, H., Su, L., Zuo, C., Shen, X., Chu, W., Li, F., & Shi, H. (2024). Rapid mass conversion for environmental microplastics of diverse shapes. *Environmental Science & Technology*, *58*(24), 10776–10785. https://doi.org/10.1021/acs.est.4c01031

Costache, R., Crăciun, A., Ciobotaru, N., & Bărbulescu, A. (2024). Intelligent methods for estimating the flood susceptibility in the Danube Delta, Romania. *Water*, *16*(23), 3511.

De La Peña, L., Guo, R., Cao, X., Ni, X., & Zhang, W. (2022). Accelerating the energy transition to achieve carbon neutrality. *Resources, Conservation and Recycling*, *177*, 105957.

Delucchi, M. A., & Jacobson, M. Z. (2011). Providing all global energy with wind, water, and solar power, Part II: Reliability, system and transmission costs, and policies. *Energy Policy*, *39*(3), 1170–1190.

Dent, S. (2023). *Recognizing Value in an Open-Architecture Digital Water System: San Jacinto River Authority's Approach to Becoming a Smart Water Utility*. WEFTEC 2023.

Du, W.-J., Lu, J.-Y., Hu, Y.-R., Xiao, J., Yang, C., Wu, J., Huang, B., Cui, S., Wang, Y., & Li, W.-W. (2023). Spatiotemporal pattern of greenhouse gas emissions in China's wastewater sector and pathways towards carbon neutrality. *Nature Water*, *1*(2), 166–175.

du Plessis, A. (2017). *Freshwater Challenges of South Africa and Its Upper Vaal River*. Springer.

du Plessis, A. (2023). Water resources from a global perspective. In *South Africa's Water Predicament: Freshwater's Unceasing Decline* (pp. 1–25). Springer.

Dwivedi, A., Agrawal, D., Jha, A., & Mathiyazhagan, K. (2023). Studying the interactions among Industry 5.0 and circular supply chain: Towards attaining sustainable development. *Computers & Industrial Engineering*, *176*, 108927.

ElFar, O. A., Chang, C.-K., Leong, H. Y., Peter, A. P., Chew, K. W., & Show, P. L. (2021). Prospects of Industry 5.0 in algae: Customization of production and new advance technology for clean bioenergy generation. *Energy Conversion and Management: X*, *10*, 100048.

Emmanuel, M., Doubleday, K., Cakir, B., Marković, M., & Hodge, B.-M. (2020). A review of power system planning and operational models for flexibility assessment in high solar energy penetration scenarios. *Solar Energy*, *210*, 169–180.

Emran, M. G. I., Barma, R., Khan, A. H., & Roy, M. (2024). Reasons behind the Water Crisis and Its Potential Health Outcomes. arXiv preprint arXiv:2403.07019.

Feria-Díaz, J. J., Correa-Mahecha, F., López-Méndez, M. C., Rodríguez-Miranda, J. P., & Barrera-Rojas, J. (2021). Recent desalination technologies by hybridization and integration with reverse osmosis: A review. *Water*, *13*(10), 1369.

Forward, T. W. (2024). Industry 5.0, quality of life, and sustainable adaptation to climate change. In K. Shukla, Y. Patil, R. Estoque, & P. López de Haro (Eds.), *Quality of Life and Climate Change: Impacts, Sustainable Adaptation, and Social-Ecological Resilience: Impacts, Sustainable Adaptation, and Social-Ecological Resilience* (p. 200). IGI Global.

Fu, G., Jin, Y., Sun, S., Yuan, Z., & Butler, D. (2022). The role of deep learning in urban water management: A critical review. *Water Research*, *223*, 118973.

Fu, G., Sun, S., Hoang, L., Yuan, Z., & Butler, D. (2023). Artificial intelligence underpins urban water infrastructure of the future: A holistic perspective. *Cambridge Prisms: Water*, *1*, e14.

Garschagen, M., & Doshi, D. (2022). Does funds-based adaptation finance reach the most vulnerable countries? *Global Environmental Change*, *73*, 102450.

Ghazi, Z. M., Rizvi, S. W. F., Shahid, W. M., Abdulhameed, A. M., Saleem, H., & Zaidi, S. J. (2022). An overview of water desalination systems integrated with renewable energy sources. *Desalination*, *542*, 116063.

Ghobakhloo, M., Iranmanesh, M., Morales, M. E., Nilashi, M., & Amran, A. (2023). Actions and approaches for enabling Industry 5.0-driven sustainable industrial transformation: A strategy roadmap. *Corporate Social Responsibility and Environmental Management*, *30*(3), 1473–1494.

Giron, D. (2019). *Singapore; brief insight into its history and urban development.* Unpublished manuscript. Academia.edu. https://www.academia.edu/38386498/Singapore_brief_insight_into_its_history_and_urban_development_pdf

Gitis, V., & Hankins, N. (2018). Water treatment chemicals: Trends and challenges. *Journal of Water Process Engineering*, *25*, 34–38.

Glied, V. (2020). Emerging Water issues in the 21st century. In *Water Stress: Emerging Challenges of Global Water Scarcity* (pp. 10–32). University of Pécs.

Goede, M. (2020). *Society 5.0; We and I.* University of Governanace/Goede Consultants.

Goel, S., & Sharma, R. (2017). Performance evaluation of stand alone, grid connected and hybrid renewable energy systems for rural application: A comparative review. *Renewable and Sustainable Energy Reviews*, *78*, 1378–1389.

Gohil, J., Patel, J., Chopra, J., Chhaya, K., Taravia, J., & Shah, M. (2021). Advent of Big Data technology in environment and water management sector. *Environmental Science and Pollution Research*, *28*(45), 64084–64102.

Golovianko, M., Terziyan, V., Branytskyi, V., & Malyk, D. (2023). Industry 4.0 vs. Industry 5.0: Co-existence, transition, or a hybrid. *Procedia Computer Science*, *217*, 102–113.

Gooden, J., & Pritzlaff, R. (2021). Dryland watershed restoration with rock detention structures: A nature-based solution to mitigate drought, erosion, flooding, and atmospheric carbon. *Frontiers in Environmental Science*, *9*, 679189.

Greco, F., Heijman, S. G., & Jarquin-Laguna, A. (2021). Integration of wind energy and desalination systems: a review study. *Processes*, *9*(12), 2181.

Gulagi, A. (2023). South Asia's energy [R]evolution–transition towards defossilised power systems by 2050 with special focus on India (Acta Universitatis Lappeenrantaensis 1067) [Doctoral dissertation, Lappeenranta-Lahti University of Technology LUT]. LUT University Repository. https://lutpub.lut.fi/handle/10024/165144

Guo, Y., Liu, Y., Georgiou, T., & Lew, M. S. (2018). A review of semantic segmentation using deep neural networks. *International Journal of Multimedia Information Retrieval*, *7*, 87–93.

Guo, Y., Liu, Y., Georgiou, T., & Lew, M. S. (2018). A review of semantic segmentation using deep neural networks. *International Journal of Multimedia Information Retrieval, 7*, 87–93.

Hansen, H. H. (2024). Predictive maintenance for power plants [Industrial doctoral dissertation, Technical University of Denmark]. DTU Research Database. https://orbit.dtu.dk/en/publications/predictive-maintenance-for-power-plants/

Harpool, S. (2018). Renewable energy in the Pacific Northwest-technical and economic analysis of 100% renewable portfolios, including seasonality impacts and potential applications of energy storage. Master's Thesis, Oregon State University. Apa Citation // ScholarsArchive@OSU

Hernández-del-Olmo, F., Gaudioso, E., Dormido, R., & Duro, N. (2018). Tackling the start-up of a reinforcement learning agent for the control of wastewater treatment plants. *Knowledge-Based Systems*, *144*, 9–15.

Hernández-del-Olmo, F., Gaudioso, E., Duro, N., & Dormido, R. (2019). Machine learning weather soft-sensor for advanced control of wastewater treatment plants. *Sensors*, *19*(14), 3139.

Hernández-del-Olmo, F., Gaudioso, E., Dormido, R., & Duro, N. (2016). Energy and environmental efficiency for the N-ammonia removal process in wastewater treatment plants by means of reinforcement learning. *Energies*, *9*(9), 755.

Horne, R., Putland, T., & Brady, M. (2023). Regulating trusted autonomous systems in Australia. arXiv preprint arXiv:2302.03778.

Hyde-Smith, L., Roelich, K., Mdee, A., Zhan, Z., & Evans, B. (2024). Blinded by the 'green-halo'? Equity in financing climate adaptation of urban sanitation. *Environment and Planning E: Nature and Space*, *8*(1), 278–305. https://doi.org/10.1177/25148486241287446

Ihsanullah, I., Atieh, M. A., Sajid, M., & Nazal, M. K. (2021). Desalination and environment: A critical analysis of impacts, mitigation strategies, and greener desalination technologies. *Science of the Total Environment*, *780*, 146585.

Ikegwu, A. C., Nweke, H. F., Anikwe, C. V., Alo, U. R., & Okonkwo, O. R. (2022). Big data analytics for data-driven industry: A review of data sources, tools, challenges, solutions, and research directions. *Cluster Computing*, *25*(5), 3343–3387.

Imandoust, M., Alghorayshi, S. T. K., Javidfar, M., Asadi, B., Jafarinasab, M., Qezelbigloo, S., & Zahedi, R. (2024). Technical-economic analysis and optimization of multiple effect distillation system by solar energy conversion as the heat source. *Solar Energy*, *280*, 112859.

Islam, M. J. (2020). Awareness of safe water, proper sanitation, hygiene practices and prevalence of water borne diseases among people of different slum areas of Chattogram City, Bangladesh. *EC Nursing and Healthcare*, *2*, 21–36.

Jacobson, M. Z., Delucchi, M. A., Cameron, M. A., & Mathiesen, B. V. (2018). Matching demand with supply at low cost in 139 countries among 20 world regions with 100% intermittent wind, water, and sunlight (WWS) for all purposes. *Renewable Energy*, *123*, 236–248.

Jain, S., Srivastava, A., Khadke, L., Chatterjee, U., & Elbeltagi, A. (2024). Global-scale water security and desertification management amidst climate change. *Environmental Science and Pollution Research*, *31*(49), 58720–58744. https://doi.org/10.1007/s11356-024-34916-0

Jayakumar, D., Bouhoula, A., & Al-Zubari, W. K. (2024). Unlocking the potential of artificial intelligence for sustainable water management focusing operational applications. *Water*, *16* (22), 3328; https://doi.org/10.3390/w16223328.

Joo, J.-G., Jeong, I.-S., & Kang, S.-H. (2024). Deep reinforcement learning for multi-objective real-time pump operation in rainwater pumping stations. *Water*, *16*(23), 3398.

Jouhara, H., Czajczyńska, D., Ghazal, H., Krzyżyńska, R., Anguilano, L., Reynolds, A., & Spencer, N. (2017). Municipal waste management systems for domestic use. *Energy*, *139*, 485–506.

Kalair, A., Abas, N., Saleem, M. S., Kalair, A. R., & Khan, N. (2021). Role of energy storage systems in energy transition from fossil fuels to renewables. *Energy Storage*, *3*(1), e135.

Kandpal, V., Jaswal, A., Santibanez Gonzalez, E. D., & Agarwal, N. (2024). Corporate social responsibility (CSR) and ESG reporting: redefining business in the twenty-first century. In *Sustainable Energy Transition: Circular Economy and Sustainable Financing for Environmental, Social and Governance (ESG) Practices* (pp. 239–272). Springer.

Kebede, A. A., Kalogiannis, T., Van Mierlo, J., & Berecibar, M. (2022). A comprehensive review of stationary energy storage devices for large scale renewable energy sources grid integration. *Renewable and Sustainable Energy Reviews*, *159*, 112213.

Khan, M. M., Siddiqi, S. A., Farooque, A. A., Iqbal, Q., Shahid, S. A., Akram, M. T., Rahman, S., Al-Busaidi, W., & Khan, I. (2022). Towards sustainable application of wastewater in agriculture: A review on reusability and risk assessment. *Agronomy*, *12*(6), 1397.

Khan, T., Yu, M., & Waseem, M. (2022). Review on recent optimization strategies for hybrid renewable energy system with hydrogen technologies: State of the art, trends and future directions. *International Journal of Hydrogen Energy*, *47*(60), 25155–25201.

Khilchevskyi, V., & Karamushka, V. (2021). Global water resources: Distribution and demand. In *Clean Water and Sanitation* (pp. 1–11). Springer.

Khondoker, M., Mandal, S., Gurav, R., & Hwang, S. (2023). Freshwater shortage, salinity increase, and global food production: A need for sustainable irrigation water desalination—A scoping review. *Earth*, *4*(2), 223–240.

Krishnan, S. R., Nallakaruppan, M., Chengoden, R., Koppu, S., Iyapparaja, M., Sadhasivam, J., & Sethuraman, S. (2022). Smart water resource management using Artificial Intelligence—A review. *Sustainability*, *14*(20), 13384.

Kumar, M. (2020). Social, economic, and environmental impacts of renewable energy resources. In K. E. Okedu, A. Tahour, & A. G. Aissaoui (Eds.), *Wind solar hybrid renewable energy system* (Chapter 1). IntechOpen. https://doi.org/10.5772/intechopen.89494

Laird, J. (2018). *Drought Water Rights Allocation Tool: Eel River Model Update and pyWRAT Application*. University of California.

Lee, J.-Y., Rahman, A., Huang, S., Smith, A. D., & Katipamula, S. (2022). On-policy learning-based deep reinforcement learning assessment for building control efficiency and stability. *Science and Technology for the Built Environment*, *28*(9), 1150–1165.

Leng, J., Sha, W., Wang, B., Zheng, P., Zhuang, C., Liu, Q., Wuest, T., Mourtzis, D., & Wang, L. (2022). Industry 5.0: Prospect and retrospect. *Journal of Manufacturing Systems*, *65*, 279–295.

Li, D., Cong, A., & Guo, S. (2019). Sewer damage detection from imbalanced CCTV inspection data using deep convolutional neural networks with hierarchical classification. *Automation in Construction*, *101*, 199–208.

Li, J., & Li, G. (2024). Natural resources utilization, digital technology, and green development expenditures can reduce environmental stress: A case study of emerging economies. *Resources Policy*, *95*, 105147.

Liang, K., Ma, Q., Lu, H., Fang, H., Yang, P., & Fan, J. (2020). The research and application progress of the isobaric ERD technique for SWRO desalination plant. *Desalination and Water Treatment*, *202*, 14–26.

Liang, R., AL- Huqail, A. A., Ali, H. E., Ponnore, J. J., Alkhalifah, T., Alturise, F., & Assilzadeh, H. (2023). Wireless water consumption sensing system for building energy efficiency: A visual-based approach with self-powered operation. *Energy and Buildings*, *301*, 113584.

Lim, Y. J., Goh, K., Kurihara, M., & Wang, R. (2021). Seawater desalination by reverse osmosis: Current development and future challenges in membrane fabrication—A review. *Journal of Membrane Science*, *629*, 119292.

Lin, S., Zhao, H., Zhu, L., He, T., Chen, S., Gao, C., & Zhang, L. (2021). Seawater desalination technology and engineering in China: A review. *Desalination*, *498*, 114728.

Liu, J., Yang, X., Lau, S., Wang, X., Luo, S., Lee, V. C. S., & Ding, L. (2020). Automated pavement crack detection and segmentation based on two-step convolutional neural network. *Computer-Aided Civil and Infrastructure Engineering*, *35*(11), 1291–1305.

Maddikunta, P. K. R., Pham, Q.-V., Prabadevi, B., Deepa, N., Dev, K., Gadekallu, T. R., Ruby, R., & Liyanage, M. (2022). Industry 5.0: A survey on enabling technologies and potential applications. *Journal of Industrial Information Integration*, *26*, 100257.

Martínez Díaz, M. d. M. (2017). Stand-alone hybrid renewable energy systems (HRES). [Doctoral dissertation, Universitat Politècnica de Catalunya]. TDX (Tesis Doctorals en Xarxa). http://hdl.handle.net/10803/457978

Martini, B., Bellisario, D., & Coletti, P. (2024). Human-centered and sustainable artificial intelligence in industry 5.0: Challenges and perspectives. *Sustainability*, *16*(13), 5448.

Martos, V., Ahmad, A., Cartujo, P., & Ordoñez, J. (2021). Ensuring agricultural sustainability through remote sensing in the era of agriculture 5.0. *Applied Sciences*, *11*(13), 5911.

Matin, A., Laoui, T., Falath, W., & Farooque, M. (2021). Fouling control in reverse osmosis for water desalination & reuse: Current practices & emerging environment-friendly technologies. *Science of the Total Environment*, *765*, 142721.

Mebar, H. (2024). *Creating an ESG services portfolio*. [Master's thesis, Haaga-Helia University of Applied Sciences]. Theseus. https://www.theseus.fi/bitstream/handle/10024/872690/Mebar_Hani.pdf

Mehrotra, N., & Benjamin, E. O. (2022). Evaluating the enhancement of the Nationally Determined Contributions (NDCs) of developing countries: an international support programme perspective. *Climate Policy*, *22*(6), 728–742.

Merry, C. J. (2001). Basic electromagnetic radiation. *Manual of Geospatial Science and Technology*, *62*, 3668–3672.

Mishra, R. K. (2023). Fresh water availability and its global challenge. *British Journal of Multidisciplinary and Advanced Studies*, *4*(3), 1–78.

Mohanty, A., Mohanty, S. K., & Mohapatra, A. G. (2024a). Real-time monitoring and fault in AI-enhanced wastewater treatment systems. In M. C. Garg (Ed.), *The AI cleanse: Transforming wastewater treatment through artificial intelligence: Harnessing data-driven solutions* (pp. 165–199). Springer. https://doi.org/10.1007/978-3-031-67237-8_7

Mohanty, A., Mohanty, S. K., & Mohapatra, A. G. (2024b). Real-time monitoring and fault detection in AI-enhanced wastewater treatment systems. In *The AI Cleanse: Transforming Wastewater Treatment through Artificial Intelligence: Harnessing Data-Driven Solutions* (pp. 165–199). Springer.

Mounce, S. R., Shepherd, W. J., Boxall, J. B., Horoshenkov, K. V., & Boyle, J. H. (2021). Autonomous robotics for water and sewer networks. *IAHR Hydrolink*, *2*, 55–62.

Mourtzis, D., Angelopoulos, J., & Panopoulos, N. (2022). A literature review of the challenges and opportunities of the transition from Industry 4.0 to Society 5.0. *Energies*, *15*(17), 6276.

Mudombi, S., & Montmasson-Clair, G. (2020). *A Case for Water and Sanitation in South Africa's Post-Lockdown Economic Recovery Stimulus Package.* Trade & Industrial Policy Strategies.

Mullapudi, A., Lewis, M. J., Gruden, C. L., & Kerkez, B. (2020). Deep reinforcement learning for the real time control of stormwater systems. *Advances in Water Resources*, *140*, 103600. https://doi.org/10.1016/j.advwatres.2020.103600

Mushtaque, M. A. R. (2024). Integration of wireless sensor networks, internet of things, artificial intelligence, and deep learning in smart agriculture: A comprehensive survey: Integration of wireless sensor networks, internet of things. *Journal of Innovative Intelligent Computing and Emerging Technologies (JIICET)*, *1*(01), 8–19.

Nandineni, R. D., & Dash, S. P. (2022). Capacity building in water resource management in developing countries. In W. Leal Filho, A. M. Azul, L. Brandli, A. Lange Salvia, & T. Wall (Eds.), *Clean Water and Sanitation* (pp. 49–61). Springer.

Negm, A., Ma, X., & Aggidis, G. (2024). Deep reinforcement learning challenges and opportunities for urban water systems. *Water Research*, *253*, 121145. https://doi.org/10.1016/j.watres.2024.121145

Nkhata, B. (2022). Climate change and water resources in southern Africa: A resilience perspective. In *The R. C. Brears (Ed.), Palgrave Handbook of Climate Resilient Societies* (pp. 413–434). Springer.

Oeij, P., Lenaerts, K., Dhondt, S., Van Dijk, W., Schartinger, D., Sorko, S., & Warhurst, C. (2024). A conceptual framework for workforce skills for Industry 5.0: Implications for research, policy and practice. *Journal of Innovation Management*, *12*(1), 205–233.

Omukuti, J., Barrett, S., White, P. C., Marchant, R., & Averchenkova, A. (2022). The green climate fund and its shortcomings in local delivery of adaptation finance. *Climate Policy*, *22*(9–10), 1225–1240.

Ortiz-Partida, J. P., Fernandez-Bou, A. S., Maskey, M., Rodríguez-Flores, J. M., Medellín-Azuara, J., Sandoval-Solis, S., Ermolieva, T., Kanavas, Z., Sahu, R. K., & Wada, Y. (2023). Hydro-economic modeling of water resources management challenges: Current applications and future directions. *Water Economics and Policy*, *9*(01), e2340003.

Oztemel, E., & Gursev, S. (2020). Literature review of Industry 4.0 and related technologies. *Journal of Intelligent Manufacturing*, *31*(1), 127–182.

Pachauri, S., Brew-Hammond, A., Barnes, D. F., Bouille, D. H., Gitonga, S., Modi, V., Prasad, G., Rath, A., & Zerrifi, H. (2012). Energy access for development. In GEA Writing Team (Eds.), *Global energy assessment: Toward a sustainable future* (Chapter 19, pp. 1401–1458). Cambridge University Press & International Institute for Applied Systems Analysis (IIASA). https://doi.org/10.1017/CBO9780511793677.023

Paroha, A. D. (2022). Integration of Internet of Things (IoT) in petroleum reservoir monitoring: A comprehensive analysis of real-time data for enhanced decision-making. *Transactions on Latest Trends in IoT*, *5*(5), 1–15.

Pawłowski, A., & Rydzewski, P. (2024). Pathways to carbon neutrality: Integrating energy strategies, policy, and public perception in the face of climate change—A global perspective. *Energies*, *17*(23), 5867.

Pereira, R., & dos Santos, N. (2023). Neoindustrialization—Reflections on a new paradigmatic approach for the industry: A scoping review on industry 5.0. *Logistics*, *7*(3), 43.

Pill, M. (2022). Towards a funding mechanism for loss and damage from climate change impacts. *Climate Risk Management*, *35*, 100391.

Pranav, M., Karunanithi, K., Akhil, M., Vanan, S. S., Afsal, V., & Krishnan, A. (2017). Hybrid renewable energy sources (HRES)—A review. *2017 International Conference on Intelligent Computing, Instrumentation and Control Technologies (ICICICT)* (pp. 1–6). IEEE. https://doi.org/10.1109/ICICICT1.2017.8342653

Qi, J., & Qian, H. (2023). Climate finance at a crossroads: It is high time to use the global solution for global problems. *Carbon Neutrality*, *2*(1), 31.

Ramirez-Meyers, K., Mann, W. N., Deetjen, T., Johnson, S., Rhodes, J., & Webber, M. (2021). How different power plant types contribute to electric grid reliability, resilience, and vulnerability: A comparative analytical framework. *Progress in Energy*, *3*(3), 033001.

Ranganathan, C. S., Vijayalakshmi, K., Kalaichelvi, S., Sivanantham, E., GaneshBabu, T., & Murugan, S. (2024). Enhancing efficiency in drain cleaning operations using k-nearest neighbors algorithm with IoT-enabled robots. In *Proceedings of the 2024 First International Conference on Innovations in Communications, Electrical and Computer Engineering (ICICEC)* (pp. 813–818). IEEE. https://doi.org/10.1109/ICICEC62498.2024.10808441

Rayhana, R., Jiao, Y., Zaji, A., & Liu, Z. (2020). Automated vision systems for condition assessment of sewer and water pipelines. *IEEE Transactions on Automation Science and Engineering*, *18*(4), 1861–1878.

Ren, S., Zhang, Y., Liu, Y., Sakao, T., Huisingh, D., & Almeida, C. M. (2019). A comprehensive review of big data analytics throughout product lifecycle to support sustainable smart manufacturing: A framework, challenges and future research directions. *Journal of Cleaner Production*, *210*, 1343–1365.

Saniuk, S., Grabowska, S., & Straka, M. (2022). Identification of social and economic expectations: Contextual reasons for the transformation process of Industry 4.0 into the Industry 5.0 concept. *Sustainability*, *14*(3), 1391.

Sarbu, I., & Popa-Albu, S. (2023). Optimization of urban water distribution networks using heuristic methods: An overview. *Water International*, *48*(1), 120–148.

Sarpong, H. A. (n.d.). Water in cop: From marginalization to centrality–the Baku water declaration and its role in shaping future climate action. Available at SSRN 5098439.

Satheesh, S., & Shukla, K. (2024). Industry 5.0, quality of life, and sustainable adaptation to climate change: Challenges and the transformational way forward. In K. Shukla, Y. B. Patil, R. C. Estoque, & P. A. López de Haro (Eds.), *Quality of Life and Climate Change: Impacts, Sustainable Adaptation, and Social-Ecological Resilience* (pp. 200–219). IGI Global.

Scheihing, K. W., Fraser, C. M., Vargas, C. R., Kukurić, N., & Lictevout, E. (2022). A review of current capacity development practice for fostering groundwater sustainability. *Groundwater for Sustainable Development*, *19*, 100823.

Serafeim, A. V., Fourniotis, N. T., Deidda, R., Kokosalakis, G., & Langousis, A. (2024). Leakages in water distribution networks: estimation methods, influential factors, and mitigation strategies—A comprehensive review. *Water*, *16*(11), 1534.

Shahid, F., & Ahmad, A. (2023). Climate change: Impacts on sustainable development goals and human rights. *Annals of Human and Social Sciences*, *4*(4), 332–341.

Shalaby, S., Sharshir, S. W., Kabeel, A., Kandeal, A., Abosheiasha, H., Abdelgaied, M., Hamed, M. H., & Yang, N. (2022). Reverse osmosis desalination systems powered by solar energy: Preheating techniques and brine disposal challenges—A detailed review. *Energy Conversion and Management*, *251*, 114971.

Sharma, R., & Gupta, H. (2024). Leveraging cognitive digital twins in industry 5.0 for achieving sustainable development goal 9: An exploration of inclusive and sustainable industrialization strategies. *Journal of Cleaner Production*, *448*, 141364.

Sharma, S., Sharma, K., & Grover, S. (2024). Real-time data analysis with smart sensors. In S. Gulati (Ed.), *Application of Artificial Intelligence in Wastewater Treatment* (pp. 127–153). Springer.

Shemer, H., Wald, S., & Semiat, R. (2023). Challenges and solutions for global water scarcity. *Membranes*, *13*(6), 612.

Shokri, A., & Fard, M. S. (2022). A sustainable approach in water desalination with the integration of renewable energy sources: Environmental engineering challenges and perspectives. *Environmental Advances*, *9*, 100281.

Smith, O., Cattell, O., Farcot, E., O'Dea, R. D., & Hopcraft, K. I. (2022). The effect of renewable energy incorporation on power grid stability and resilience. *Science Advances*, *8*(9), eabj6734.

Snelgrove, J. P., Tharrett, D. L., Pittaluga, A. E., & Decker, J. K. (2020). Robotic autonomous systems: Manned/unmanned teaming (*RAS-MUM-T*) [Graduate research project report, Naval Postgraduate School]. *The NPS Institutional Archive; Defense Technical Information Center (DTIC)*. https://calhoun.nps.edu/handle/10945/66732

Song, C., Zhu, J.-J., Yuan, Z., van Loosdrecht, M. C., & Ren, Z. J. (2024). Defining and achieving net-zero emissions in the wastewater sector. *Nature Water*, *2*(10), 927–935. https://doi.org/10.1038/s44221-024-00318-2

Spadoro, J., Vombergar, K., & Hoempler, M. (2024). Flying high: The role of drones in enhancing utility services. *Climate and Energy*, *40*(9), 1–12.

Spellman, F. R. (2013). *Water & Wastewater Infrastructure: Energy Efficiency and Sustainability*. CRC Press.

Swedenborg, W., Timboe, I., Hebart-Coleman, D., Kjellen, M., Matthews, J., Davis, K., Stein, N., & Sato, I. (2022). *NDC Enhancement: Opportunities Through Water*. World Resource Institute, US. Report.

Taboada, J. V., Diaz-Casas, V., & Yu, X. (2021). Reliability and maintenance management analysis on offshore wind turbines (OWTs). *Energies*, *14*(22), 7662.

Tai, H., Samandari, H., Pacthod, D., Polymeneas, E., & Bolano, A. (2022). *The energy transition: A region-by-region agenda for near-term action*. McKinsey & Company. https://www.mckinsey.com/industries/electric-power-and-natural-gas/our-insights/the-energy-transition-a-region-by-region-agenda-for-near-term-action

Teisho, O. (2022). *Energy and Development: An Assessment of South Africa's progress towards achieving SDG 7*. Department of Anthropology and Archaeology, Faculty of Humanities [Master's dissertation, University of Pretoria]. University of Pretoria Institutional Repository. https://repository.up.ac.za/handle/2263/89504

Therrien, J.-D., Nicolaï, N., & Vanrolleghem, P. A. (2020). A critical review of the data pipeline: How wastewater system operation flows from data to intelligence. *Water Science and Technology*, *82*(12), 2613–2634.

Thongtam, N., Sinthupinyo, S., & Chandrachai, A. (2024). Optimization of intelligent maintenance system in smart factory using state space search algorithm. *Applied Sciences*, *14*(24), 11973.

Tipon Tanchangya, A. R., Rahman, J., & Ridwan, M. (2024). A review of deep learning applications for sustainable water resource management. *Global Sustainability Research*, *3*(4), 48–73. https://doi.org/10.56556/gssr.v3i4.1043

United Nations Educational, Scientific and Cultural Organization (UNESCO). (2021). *The United Nations World Water Development Report 2021: Valuing water*. UNESCO. https://www.unwater.org/publications/un-world-water-development-report-2021

United Nations Educational, Scientific and Cultural Organization (UNESCO). (2019). *The United Nations World Water Development Report 2019: Leaving no one behind*. UNESCO. https://www.unwater.org/publications/un-world-water-development-report-2019

Vahidhosseini, S. M., Rashidi, S., Hsu, S.-H., Yan, W.-M., & Rashidi, A. (2024). Integration of solar thermal collectors and heat pumps with thermal energy storage systems for building energy demand reduction: A comprehensive review. *Journal of Energy Storage*, *95*, 112568.

Vakilifard, N., Anda, M., Bahri, P. A., & Ho, G. (2018). The role of water-energy nexus in optimising water supply systems—Review of techniques and approaches. *Renewable and Sustainable Energy Reviews*, *82*, 1424–1432.

Viollet, P.-L. (2007). *Water Engineering in Ancient Civilizations: 5,000 Years of History*. CRC Press.

Wang, M., & Cheng, J. C. P. (2018). Development and improvement of deep learning based automated defect detection for sewer pipe inspection using Faster R-CNN. In I. Smith & B. Domer (Eds.), *Advanced computing strategies for engineering: EG-ICE 2018* (Lecture Notes in Computer Science, vol. 10864, pp. 171–192). Springer. https://doi.org/10.1007/978-3-319-91638-5_9

Wałachowska, A., & Ignasiak-Szulc, A. (2021). Comparison of renewable energy sources in 'New' EU Member States in the context of national energy transformations. *Energies*, *14*(23), 7963.

Walter, C., Saenz, J., Elkmann, N., Althoff, H., Kutzner, S., & Stuerze, T. (2012). Design considerations of robotic system for cleaning and inspection of large-diameter sewers. *Journal of Field Robotics*, *29*(1), 186–214.

Whittington, C. (2016). *Russian River Drought Water Right Allocation Tool (DWRAT)*. University of California.

World Health Organization. (2021). *State of the World's Hand Hygiene: A Global Call to Action to Make Hand Hygiene a Priority in Policy and Practice*. World Health Organization.

Xu, L. D., Xu, E. L., & Li, L. (2018). Industry 4.0: State of the art and future trends. *International Journal of Production Research*, *56*(8), 2941–2962.

Yang, J., Liu, J., Xiang, Y., Zhang, S., & Liu, J. (2022). Data-driven optimal dynamic dispatch for Hydro-PV-PHS integrated power systems using deep reinforcement learning approach. *CSEE Journal of Power and Energy Systems*, *9*(3), 846–858.

Yaseen, A. (2024). Enhancing cybersecurity through automated infrastructure management: A comprehensive study on optimizing security measures. *Quarterly Journal of Emerging Technologies and Innovations*, *9*(1), 38–60.

Yasin, H. Q., Breadsell, J., & Tahir, M. N. (2021). Climate-water governance: A systematic analysis of Pakistan's water sector resilience and adaptation to combat climate change. *Water Policy*, *23*(1), 1–35.

Yatawatta, Y. J. M., & Sridarran, P. (2024). Investigating the barriers and strategies for establishing desalination plants to mitigate water scarcity in Sri Lankan dry zones. *Sustainability Accounting, Management and Policy Journal*, *15*(6), 1408–1441.

Yu, Y., Appiah, D., Zulu, B., & Adu-Poku, K. A. (2024). Integrating rural development, education, and management: Challenges and strategies. *Sustainability, 16*(15), 6474.

Yuan, J., Mao, W., Hu, C., Zheng, J., Zheng, D., & Yang, Y. (2023). Leak detection and localization techniques in oil and gas pipeline: A bibliometric and systematic review. *Engineering Failure Analysis, 146*, 107060.

Zaman, D., Tiwari, M. K., Gupta, A. K., & Sen, D. (2020). A review of leakage detection strategies for pressurised pipeline in steady-state. *Engineering Failure Analysis, 109*, 104264.

Zapata-Sierra, A., Cascajares, M., Alcayde, A., & Manzano-Agugliaro, F. (2021). Worldwide research trends on desalination. *Desalination, 519*, 115305.

Zhang, Y. (2024). A path to sustainable development of agri-industries: Analysis of agriculture 5.0 versus industry 5.0 using stakeholder theory with moderation of environmental policy. *Sustainable Development, 32*(5), 4829–4843. https://doi.org/10.1002/sd.2923

Zhang, Y.-F., & Thorburn, P. J. (2022). A deep surrogate model with spatio-temporal awareness for water quality sensor measurement. *Expert Systems with Applications, 200*, 116914.

10 Cybersecurity and Risk Management in Water Systems

10.1 INTRODUCTION TO CYBERSECURITY AND RISK MANAGEMENT IN WATER SYSTEMS

Incorporating cutting-edge technologies like big data analytics, machine learning (ML), artificial intelligence (AI), and the Internet of Things (IoT) defines the shift of water systems into Water 4.0 and 5.0 environments. Although these technologies improve predictive maintenance, resource management, and operational efficiency, they pose serious cybersecurity threats. According to the Cybersecurity and Infrastructure Security Agency (CISA), water systems are a critical infrastructure sector most commonly targeted by cyberattacks, with ransomware instances rising sharply in recent years (Moraitis et al., 2023). Additionally, according to a 2023 poll by the Water Information Sharing and Analysis Center (WaterISAC), 68% of North American water utilities had at least one cybersecurity incident within the previous three years, with the average cost of such an attack estimated at $3.86 million, including operational downtime and remediation costs (Tuptuk et al., 2021).

The weaknesses in network infrastructures, Supervisory Control and Data Acquisition (SCADA) systems, and IoT devices are particularly worrisome. IoT equipment, such as smart meters and water quality sensors, frequently have low processing power, so putting strong encryption and authentication procedures in place can be complicated. Since many of these devices use default credentials or out-of-date firmware, they are vulnerable to botnet infections and brute-force attacks. According to research, a sizable portion of IoT devices in critical infrastructure domains, such as water systems, are still susceptible to cyberattacks because of inadequate security measures (Tuptuk et al., 2021).

SCADA systems are crucial for monitoring and managing water processes but were not initially built with cybersecurity. They frequently run on outdated systems that are incompatible with contemporary security protocols, leaving them open to attacks such as denial-of-service and man-in-the-middle attacks (Tuptuk et al., 2021; Moraitis et al., 2023).

Advanced Persistent Threats, malware, and phishing attacks frequently target network infrastructures that link cloud platforms, SCADA systems, and IoT devices. The lack of intrusion detection systems (IDS) and inadequate network segmentation enable attackers to move laterally within the system, increasing their privileges and obtaining access to vital components. The urgent need for improved security measures is shown by over 40% of water utility experts who rank ransomware as their

DOI: 10.1201/9781003633839-10

top cybersecurity threat (Tuptuk et al., 2021). Despite the advancements in water system security, gaps remain. The specificity required to handle the particular needs of water systems functioning in Water 4.0 and 5.0 contexts is frequently lacking in cybersecurity frameworks (Nikolopoulos et al., 2021; Tuptuk et al., 2021). Promising solutions are offered by emerging trends, such as using AI and ML algorithms to analyse network traffic patterns and identify anomalies instantly, putting Zero Trust Architecture into practice to authenticate each user and device inside the network perimeter, and investigating blockchain technology to protect data exchange between SCADA systems and IoT devices. As frameworks are being created to integrate cybersecurity measures with physical security systems, forming a holistic defence mechanism, cyber-physical security integration is also becoming more popular (Nikolopoulos et al., 2021; Tuptuk et al., 2021). These innovations are critical to addressing the growing vulnerabilities and ensuring the resilience of critical water infrastructure against evolving cyber threats.

10.2 DEVELOPING END-TO-END CYBERSECURITY POLICIES FOR WATER UTILITIES

Water utilities must have strong end-to-end cybersecurity policies in place to protect vital infrastructure from a growing number of cyber threats. The attack surface grows considerably as water utilities incorporate more and more cutting-edge technology, including cloud-based platforms, SCADA systems, and IoT devices. Extensive, multi-layered security procedures are required to safeguard these systems against potential weaknesses and attacks (Clark et al., 2017; Hassanzadeh et al., 2020; You, 2022).

Segmenting a network into several zones can significantly lower the possibility of attackers moving laterally. Monitoring and managing traffic between these segments require the implementation of sophisticated firewalls, IDS, and intrusion prevention systems (Pliatsios et al., 2020; Wai, 2023). Secure communication protocols like end-to-end encryption and Virtual Private Networks should be implemented to safeguard data further while it is being transmitted. Strict security guidelines, such as frequent firmware updates, robust authentication procedures, and encrypted data transmission, must be followed by IoT devices used by water utilities. Policies should require that default credentials be changed immediately after deployment and ensure that devices are regularly checked for irregularities (Hassanzadeh et al., 2020; Wai, 2023; Wali, 2024).

Layered security measures, including firewalls, secure remote access protocols, and physical access controls, are necessary for SCADA systems. It's also crucial to update or modify outdated SCADA systems to meet contemporary cybersecurity standards (Stouffer et al., 2015; Naidoo & Jacobs, 2023). Cyber incident detection, containment, and mitigation require a thorough incident response plan (IRP). Clear roles and duties, communication guidelines, and escalation methods should all be outlined in this strategy. To guarantee quick service restoration, recovery plans need to incorporate regular backups, redundant systems, and disaster recovery testing (Stouffer et al., 2015; Wai, 2023). Since human error continues to be a major contributor to cyber events, frequent training programmes should teach staff members about social engineering, phishing, and cybersecurity best practices. All organisational levels must have a security-aware culture fostered by policies (Clark et al., 2017; Hassanzadeh et al., 2020; You, 2022). It is also critical to adhere to pertinent cybersecurity laws and guidelines, such as those set forth by the National Institute of Standards and Technology (NIST) and the CISA. Regular audits and reports must be carried out to show compliance and pinpoint areas in need of development (Stouffer et al., 2015; Caithamlová, 2024). By facilitating real-time threat detection and predictive analytics, emerging technologies like AI and ML can improve cybersecurity safeguards. Blockchain technology can also be investigated for tamper-proof record-keeping and secure data integrity (Derhab et al., 2019; Hassanzadeh et al., 2020). Water utilities must actively cooperate with trade associations, governmental bodies, and cybersecurity groups to exchange threat intelligence and best practices. Collective resilience against cyber threats can be improved by participating in information-sharing programmes like the WaterISAC (Clark et al., 2017; Hassanzadeh et al., 2020; You, 2022).

Water utilities may address present vulnerabilities and foresee future difficulties by creating and implementing these thorough cybersecurity strategies. In increasingly digitalised contexts, utilities may guarantee water services' sustainability, security, and dependability by taking a proactive, end-to-end strategy (Clark et al., 2017; Hassanzadeh et al., 2020; You, 2022).

10.3 THE APPLICATION OF SOME CRITICAL TECHNOLOGIES: ENCRYPTION, FIREWALLS, MULTI-FACTOR AUTHENTICATION, AND NETWORK SEGMENTATION

It is impossible to overestimate the importance of cybersecurity in safeguarding vital infrastructure, like water utilities, in today's more interconnected world. Decisive steps are required to protect sensitive data and guarantee operational continuity due to increased cyber threats. In this effort, technologies like network segmentation, firewalls, multi-factor authentication (MFA), and encryption are essential. By guaranteeing that vital systems continue to function and be safe, these technologies protect against changing threats and contribute to the preservation of stakeholder trust (Clark et al., 2018; You, 2022).

Encryption is a key technology for protecting data in transit and at rest. Encryption prevents unwanted access to sensitive data, such as customer records and water system characteristics, by using cryptographic techniques to convert readable data into unreadable data. Encryption is crucial for protecting communications between cloud platforms, SCADA systems, and IoT devices in the context of water utilities. This keeps hackers from intercepting or altering data (Clark et al., 2018; You, 2022). Additionally, encryption helps with legal compliance, especially in settings where data privacy rules require that operational and personal data be protected (Clark et al., 2018; You, 2022).

By monitoring and managing network traffic according to pre-established security rules, firewalls serve as the first defence against outside threats. They are essential for blocking unwanted access and protecting against harmful activity like viruses and cyberattacks. Water utilities can use various firewall types, such as network and application firewalls, to safeguard SCADA systems and IoT devices. Firewalls improve the security posture of critical infrastructure by screening incoming traffic to ensure that only valid operational data is permitted (Clark et al., 2018; You, 2022). By monitoring traffic patterns and thwarting connection-based attacks, stateful firewalls—which monitor the status of active connections—offer extra protection (Clark et al., 2018; You, 2022).

By prompting users to present many forms of identity before accessing vital systems, MFA improves access restrictions. Usually, this consists of the user's knowledge (such as a password), possessions (such as a one-time code), and identity (biometric information). MFA considerably lowers the possibility of unwanted access, even if a user's password is stolen. Since a single compromised account could cause extensive operational interruptions, MFA is especially crucial for water utilities in securing administrative access to SCADA systems and databases (Clark et al., 2018; You, 2022). MFA improves overall cybersecurity resilience by guaranteeing that only authorised individuals can access sensitive control systems (Clark et al., 2018; You, 2022). Network segmentation is breaking up a network into discrete, smaller parts, each with its security measures. Restricting attackers' lateral mobility within the system lowers the possibility of extensive harm and contains possible breaches. Water utilities, for example, can divide their networks into segments to separate SCADA systems from administrative networks or IoT devices. This way, a breach in one segment will not affect the system as a whole (Clark et al., 2018; You, 2022). Furthermore, segmentation improves the security framework by enabling more detailed monitoring and identifying questionable activity (Clark et al., 2018; You, 2022).

The ability to quickly identify and address cybersecurity incidents is further improved by incorporating real-time analytics and sophisticated monitoring systems. While security information and event management (SIEM) platforms compile and examine security data from many sources to spot patterns and possible threats, real-time IDS watch network traffic for indications of malicious behaviour (Clark et al., 2018; You, 2022). Real-time detection of suspicious activity is essential for water utilities because delays in reaction can cause significant interruptions to operations (Clark et al., 2018; You, 2022).

ML and AI are being used more and more to forecast possible vulnerabilities and enhance the identification of new threats. These technologies can analyse large volumes of data from network traffic and security logs to find trends and spot irregularities that could be signs of cyberattacks. Water utilities can react more proactively to cybersecurity problems by using AI and ML to automate threat detection and analysis (Adelani, 2024a, 2024b; Labu, 2024; Onih, 2024). Water utilities are facing increasing cyber threats, and integrating these technologies into a complete cybersecurity framework provides a multi-layered defence against these attacks (Adelani, 2024a, 2024b; Labu, 2024; Onih, 2024). Water utilities may establish a strong cybersecurity posture by combining AI and ML with encryption, firewalls, MFA, network segmentation, and advanced monitoring tools. Together, these technologies guarantee business continuity and safeguard vital assets, making them crucial investments in the face of changing cyber threats (Clark et al., 2018; You, 2022).

10.4 THE ROLE OF AI IN PROACTIVE CYBERSECURITY MONITORING AND RESPONSE

AI has become a game-changing technology in cybersecurity, especially in proactive monitoring and response tactics. The number, variety, and pace of contemporary cyber threats make traditional cybersecurity measures inadequate, which frequently rely on preset regulations and manual oversight. The security of vital systems like water utilities and other vital infrastructure is greatly improved by AI technologies, such as ML, natural language processing, and predictive analytics, which offer sophisticated capabilities to detect and eliminate threats in real time (Rangaraju, 2023; Camacho, 2024; Zohuri, 2024).

The capacity of AI to evaluate enormous volumes of data at previously unheard-of rates is one of its most important benefits in cybersecurity. Monitoring logs, network traffic, and device actions in typical cybersecurity setups create large data sets that are not feasible for manual analysis. AI-powered systems can process and correlate these data points and hint at potential cyber threats. AI, for example, can identify anomalous network traffic, unexpected data transfers, or odd login habits and flag them as possible security issues before they become more serious (Kumar et al., 2023; Rizvi, 2023). According to a 2023 study, companies that used AI-driven security systems saw a notable decrease in incident detection and response times compared to conventional techniques, highlighting AI's effectiveness in improving operational efficiency (Kumar et al., 2023).

AI's involvement in cybersecurity is further strengthened by its predictive capabilities. By examining past attack data, ML systems can spot patterns that indicate new dangers. Because of its predictive power, organisations can take proactive steps to reduce hazards before they become real, such as improving access restrictions or upgrading firewall rules. AI systems, for instance, can identify malware types that target related businesses and proactively modify defence to thwart those attacks (Shahana, 2024). This pre-emptive approach is critical for water utilities, where hackers can impair vital services and pose serious hazards to public safety (Camacho, 2024).

AI improves the accuracy and timeliness of containment measures in incident response. AI-powered automated response systems may instantly revoke access rights, isolate compromised devices, and block malicious IP addresses, significantly reducing the time needed to eliminate threats and limit possible harm. According to a report, companies using AI for incident response saved significant sums of money per breach because of the quicker containment and cleanup (Kumar et al., 2023; Rizvi, 2023). This demonstrates how using AI in cybersecurity systems has economic advantages.

Additionally, through ongoing learning, AI strengthens cybersecurity resilience overall. Unlike static rule-based systems, AI algorithms react to the constantly shifting danger landscape by learning from fresh data. Deep learning models, for example, are instrumental in thwarting zero-day attacks since they can identify previously undiscovered attack patterns. Furthermore, AI can search threat intelligence feeds, dark web forums, and open-source intelligence, thanks to its natural language processing skills, which reveal novel attack methods and vulnerabilities (Karunamurthy, 2023; Shoetan, 2024).

Additionally, AI is essential for reducing cybersecurity teams' workload. Because traditional systems produce so many false positives, cybersecurity experts frequently suffer from alert fatigue. AI solves this problem by giving teams actionable insights and prioritising warnings according to severity, allowing them to focus on real risks. This efficiency is especially advantageous for firms with low resources, like smaller water utilities, where cybersecurity knowledge may be lacking (Mmaduekwe, 2024).

The application of AI in cybersecurity is not without difficulties despite its many benefits. The quality of the data used to train AI models determines their efficacy; biased or insufficient data sets may result in false threat detection. Additionally, there is rising concern about adversarial AI techniques, in which attackers alter data to avoid detection. Organisations must take a holistic strategy to address these issues, which includes integrating human oversight to verify AI-driven decisions, updating AI models often with fresh threat intelligence, and rigorously training AI models (Sinha, 2023; Mmaduekwe, 2024).

AI revolutionises proactive cybersecurity monitoring and response by allowing businesses to identify, anticipate, and neutralise attacks with previously unheard-of speed and precision. AI is a vital tool for protecting vital infrastructure from the constantly changing cyber threat scenario because of its capacity to evaluate enormous volumes of data, anticipate new threats, and automate responses. Organisations hoping to avoid more complex cyber attackers must incorporate AI technology into cybersecurity frameworks as they develop (Rangaraju, 2023; Temara, 2024; Zohuri, 2024).

10.5 GLOBAL CYBERSECURITY STANDARDS AND REGULATIONS FOR WATER SYSTEMS

Strong international cybersecurity standards and regulations are now essential as water systems embrace digital technologies like cloud-based platforms, SCADA systems, and IoT devices. By reducing the risks associated with cyberattacks, these guidelines are intended to preserve vital infrastructure, guarantee business continuity, and protect public health. Water utilities must implement comprehensive frameworks to improve their cybersecurity resilience because their reliance on digital technologies exposes them to cyberattacks (Adelani, 2024a, 2024b).

In the United States, the NIST provides a well-known cybersecurity architecture that water utilities and other critical infrastructure sectors can use as guidance. Identify, Protect, Detect, Respond, and Recover are the five main tasks in the NIST Cybersecurity Framework. To reduce cyber risks, this framework strongly emphasises risk management, asset identification, and the application of strong security controls. Water utilities can use this approach to prioritise risks, evaluate vulnerabilities, and create IRPs (Taherdoost, 2022). Furthermore, the CISA offers sector-specific instructions. For example, the "Water and Wastewater Systems Cybersecurity Framework Implementation Guide" includes practical suggestions for improving cybersecurity posture (Adelani, 2024a, 2024b).

Water utilities and other operators of critical services are subject to cybersecurity regulations set forth by the Directive on Security of Network and Information Systems (NIS Directive) in Europe. This directive requires member states to identify vital water supply organisations and make sure that appropriate security measures are put in place. It also encourages cooperation between public and commercial entities to exchange threat intelligence, strengthening critical infrastructure resilience. According to recent statistics from the European Union Agency for Cybersecurity, cyberattacks targeting the energy and water sectors have significantly increased, underscoring the significance of adhering to the NIS Directive.

A globally accepted standard for information security management systems is ISO/IEC 27001, created by the International Organization for Standardization (ISO). Water utilities that want to implement thorough security measures to safeguard private information and maintain system integrity can use this standard. To help utilities adjust to changing threats, ISO/IEC 27001 strongly emphasises risk assessment, incident response planning, and continuous improvement (Taherdoost, 2022). Businesses that have earned this certification show that they are dedicated to cybersecurity best practices, improving their standing and gaining stakeholders' trust (Adelani, 2024a, 2024b).

Even though cybersecurity standards have been established, there are still issues with their worldwide harmonisation. Regionally disparate standards lead to inconsistent application, leaving holes for hackers to exploit. Furthermore, lacking funding and technical know-how makes it difficult for many smaller water utilities to comply with regulations. According to a World Economic Forum assessment, only a small percentage of water utilities globally claimed complete compliance with cybersecurity standards, highlighting the urgent need for more funding and capacity-building initiatives (Adelani, 2024a, 2024b).

A growing emphasis on real-time threat detection and information exchange is one of the emerging trends in worldwide cybersecurity regulation for water systems. The collective defence against cyberattacks is strengthened by organisations like the WaterISAC, which gives utilities a place to exchange threat intelligence and best practices (Adelani, 2024a, 2024b). To improve the cybersecurity posture of water systems worldwide, such cooperative frameworks must be established.

In order to safeguard vital infrastructure and guarantee the security and dependability of the water supply, international cybersecurity guidelines and laws for water systems are essential. Although frameworks such as the NIST Cybersecurity Framework, ISO/IEC 27001, and the NIS Directive offer helpful guidelines, more work is required to standardise these standards internationally and facilitate compliance among utilities with limited resources. Adherence to these guidelines and proactive cooperation among stakeholders will be essential in protecting one of the most important resources in the world as cyber threats targeting water systems continue to increase.

10.6 BEST PRACTICES FOR COMPLIANCE AND SECURE DATA MANAGEMENT

Organisations, especially those in charge of vital infrastructure like water systems, must ensure that cybersecurity laws are followed and safe data management procedures are implemented. Because cyberattacks are becoming more sophisticated, enterprises must follow best practices that not only guarantee regulatory compliance but also strengthen their overall security posture; protecting sensitive data, preserving system integrity, and reducing the risk of breaches that could interfere with vital services depending on effective data management procedures (Malatji & Solms, 2021).

Establishing a strong governance structure is the first step towards achieving compliance and secure enterprise data management. This entails determining which laws—among other industry-specific standards—apply, such as ISO/IEC 27001, the NIST Cybersecurity Framework, and the General Data Protection Regulation. A specialised compliance team should map data flows, evaluate organisational procedures, and spot areas where regulations are not met. According to research, companies that closely monitor compliance gaps can significantly lower the risk of penalties and violations (Nair, 2023).

A key component of safe data management is data encryption. Organisations should use end-to-end encryption to safeguard data while it is in transit and at rest, guaranteeing that private information is kept private even if it is intercepted. Data exchanges should be secured using Transport Layer Security protocols and Advanced Encryption Standards, especially in SCADA networks and IoT-enabled water systems (Malatji & Solms, 2021). Encryption keys must be managed securely, with only authorised workers having restricted access.

Access control measures must also be implemented to guarantee that only authorised users can access sensitive systems and data. Improving identity verification procedures involves putting MFA and role-based access control into place. Regular access audits are crucial to ensure that user permissions match job duties. According to an IBM study, unauthorised access is a major contributor to data breaches, highlighting the urgent need for strong access controls (Malatji & Solms, 2021).

For disaster recovery and safe data management, regular data backups are essential. Businesses should set up automated backup systems that save copies of important data in the cloud or safe, off-site locations. Backups must be encrypted and checked regularly to guarantee data integrity and the capacity to promptly restore operations in the case of a ransomware attack or system failure. According to industry statistics, businesses can drastically save downtime during cybersecurity disasters by implementing efficient backup and recovery plans (Malatji & Solms, 2021).

Another crucial step is to put data reduction strategies into action. In order to minimise the attack surface and limit the exposure of sensitive information in the event of a breach, organisations should only gather, process, and keep the data required for operational and regulatory purposes. Data retention policies should align with compliance standards, guaranteeing that superfluous or out-of-date information is anonymised or safely erased (Malek, 2021).

Monitoring and logging are essential for data security and compliance. Organisations should implement sophisticated monitoring technologies like SIEM systems to track access and changes to sensitive data in real time. By offering consolidated insight into network activity, these technologies make it possible to identify questionable activities and streamline incident response. According to reports, businesses using real-time monitoring tools can identify and address dangers better than those using only manual procedures (Malatji & Solms, 2021).

Another essential element of safe data management is employee training. Employee education regarding best practices, like identifying phishing efforts, creating strong passwords, and securely handling sensitive data, is essential because human error is still one of the leading causes of data breaches. The probability of successful attacks can be considerably decreased by regular training sessions and simulated phishing exercises (Nair, 2023). Conduct security audits and compliance evaluations regularly to promote a continuous development culture. These audits aid in finding weaknesses, assessing how healthy controls work, and ensuring that changing legal requirements are followed. The organisation's security posture can be further improved by collaborating with outside security specialists to conduct periodic penetration tests (Malatji & Solms, 2021).

Organisations can guarantee compliance and secure data management by implementing these practices: creating governance frameworks, encrypting data, enforcing access controls, ensuring regular backups, minimising data collection, deploying real-time monitoring tools, educating staff, and conducting audits. These steps are essential for protecting sensitive data and preserving operational resilience at a time when data breaches cost an average of $4.45 million worldwide (Malatji & Solms, 2021).

10.7 THE IMPORTANCE OF MULTI-STAKEHOLDER COLLABORATION IN CYBERSECURITY

It is becoming more widely acknowledged that cybersecurity is a complex issue that cannot be successfully handled by one organisation acting alone. A cooperative strategy that includes various stakeholders, including governments, corporate organisations, industry specialists, academics, and the general public, is necessary due to the complexity and dynamic nature of cyber threats. In order to create comprehensive strategies that improve collective resilience against cyber threats, promote knowledge sharing, and guarantee that cybersecurity initiatives are flexible enough to respond to new threats, multi-stakeholder collaboration is crucial (Gordon et al., 2018; Falch et al., 2023; Obi, 2024).

One significant benefit of this cooperative strategy is the pooling of resources and expertise from many sectors. While private firms provide technical innovation and real-world applications, governments play a critical role by providing money, intelligence, and regulatory frameworks. Public advocacy organisations ensure that cybersecurity measures are fair and privacy-conscious, and academia is essential for research insights and talent development (Rantos et al., 2020; Familoni, 2024). In the United States, for example, the CISA encourages public-private partnerships through programmes such as the Joint Cyber Defense Collaborative, which seeks to improve information sharing among stakeholders, threat detection, and incident response (Gordon et al., 2018; Falch et al., 2023).

Real-time threat intelligence sharing is one of the most important advantages of multi-stakeholder collaboration. Timely access to information is crucial for effective defensive measures due to the rapid appearance and spread of cyber threats. Platforms like ISACs make sharing best practices, vulnerabilities, and threat intelligence easier for member businesses. Organisations taking part in threat intelligence-sharing programmes have dramatically decreased the average cost of a cyber intrusion, according to a World Economic Forum analysis, highlighting the observable advantages of cooperation (Falch et al., 2023).

Additionally, multi-stakeholder cooperation promotes the creation of standardised cybersecurity frameworks and procedures. The NIST Cybersecurity Framework and ISO/IEC 27001 are two standards frequently resulting from lengthy stakeholder consultation and consensus-building. These frameworks guarantee a uniform strategy for risk management, system security, and sensitive data protection across different sectors and geographical areas (Rantos et al., 2020; Obi, 2024).

A coordinated cybersecurity plan must also include personnel development and capacity growth. In order to teach and upskill the next generation of cybersecurity experts, academia, industry, and governments must work together to address the worldwide cybersecurity talent gap, which is expected to reach 3.4 million open positions by 2024 (Familoni, 2024). To close this skills gap and promote innovation, cooperative activities like cybersecurity contests and public-private apprenticeship programmes are essential (Rantos et al., 2020; Obi, 2024).

Because cyberattacks can have a cascading effect on several businesses, industries, or even countries, coordination is crucial for incident response and recovery. Minimised coordinated activities make a quick, coordinated response to minimise damage and resume operations possible. The 2017 Not Petya ransomware attack exemplifies how crucial international cooperation is when dealing with significant cyber events (Rantos et al., 2020; Obi, 2024).

Notwithstanding the evident benefits of multi-stakeholder cooperation, obstacles exist, such as conflicting priorities, communication difficulties, and data security and privacy concerns. Building explicit frameworks for data sharing, accountability, and decision-making is essential to overcoming these obstacles, as is stakeholder trust (Rantos et al., 2020; Familoni, 2024). Governments and business executives need to make investments in establishing connections and systems that allow for safe cooperation while upholding privacy and legal requirements.

Collaboration across multiple stakeholders is essential to building a strong cybersecurity ecosystem. Stakeholders improve their ability to predict, protect against, and recover from cyberattacks by exchanging intelligence, standardising procedures, addressing skills shortages, and coordinating responses. Collaboration is no longer just a best practice but a strategic need as cyber dangers continue to rise, putting companies that participate in these collaborations in a better position to protect their operations, data, and systems (Gordon et al., 2018; Falch et al., 2023; Obi, 2024).

10.8 FRAMEWORKS FOR RAPID RESPONSE AND INCIDENT RECOVERY IN THE EVENT OF CYBERATTACKS

Organisations must have strong frameworks for incident recovery and quick response as cyberattacks become more common and sophisticated to reduce downtime, lessen damage, and maintain business continuity. From discovery to recovery, these frameworks offer an organised method for handling cyber disasters, allowing firms to react efficiently and preserve stakeholder trust. Organisations with well-tested IRPs save an average of $2.66 million in breach-related expenses compared to those without, according to a 2023 IBM analysis, underscoring the vital significance of readiness. A well-designed incident response and recovery framework typically consists of the following phases, as shown in Figure 10.1.

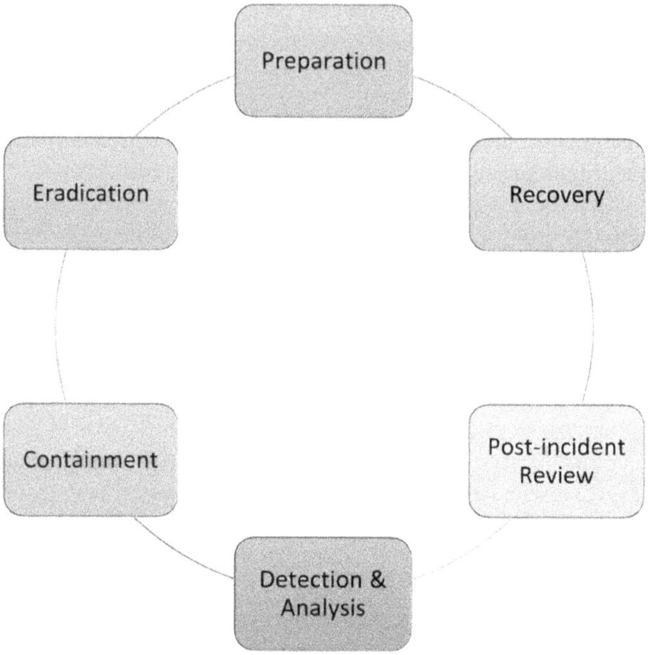

FIGURE 10.1 Incident response and recovery framework illustrating the cyclical process from preparation through detection and analysis, containment, eradication, recovery, and post-incident review.

10.8.1 Cyberattacks

Organisations must give the creation and frequent revision of their IRP top priority while creating an efficient incident response system. The roles, duties, and processes required during a cybersecurity event are outlined in this plan, which acts as a fundamental document. Forming an Incident Response Team (IRT), necessary for organising the response activities, is a crucial part of this architecture, highlighting the importance of regular cybersecurity training for team members to guarantee readiness (Alothman et al., 2022; Kanthimathinathan, 2023). Additionally, assessing the IRP and improving the team's preparedness to react efficiently requires executing tabletop exercises that mimic different assault situations (Seker & Ozbenli, 2018; Chockalingam & Maathuis, 2022). Connecting with other parties, such as law enforcement and cybersecurity companies, is essential for promoting cooperation during crises. In times of crisis, these collaborations can offer crucial extra resources and knowledge (Onwubiko & Ouazzane, 2019). Reducing the impact of attacks requires early detection of cyber threats. Advanced monitoring technologies such as IDSs, SIEM platforms, and Endpoint Detection and Response solutions are recommended for deployment by organisations. By detecting abnormalities in user behaviour and network traffic, these technologies allow quick detection and action (Vielberth et al., 2020; Zakroum et al., 2022). The average dwell period for cyber incidents worldwide is roughly 21 days, according to Mandiant research, highlighting the significance of quick detection and response systems (Vielberth et al., 2020).

To choose a suitable response plan, the IRT must assess the attack's breadth and type after it has been identified. In order to isolate compromised systems and stop the assault from spreading further, effective containment techniques are crucial. As part of their containment efforts, organisations can stop affected devices, revoke compromised credentials, and partition their networks (Carias et al., 2020a, 2020b; Kanthimathinathan, 2023). In order to prevent repeat occurrences, it is important to distinguish between short-term and long-term containment efforts. While immediate activities aim to stop the danger, long-term tactics include fixing vulnerabilities and improving security controls (Carias et al., 2020a, 2020b). For example, quickly isolating compromised systems during ransomware attacks can protect backups and important assets from encryption (Carias et al., 2020a, 2020b; Alothman et al., 2022).

Organisations must determine and eliminate the attack's primary cause after containment. To stop recurrence, this procedure may entail eliminating malware, patching exploited vulnerabilities, and strengthening security measures (Chockalingam & Maathuis, 2022; Kanthimathinathan, 2023). A thorough forensic study is frequently carried out to obtain proof and comprehend the attacker's techniques. This information can guide future security enhancements (Carias et al., 2020a, 2020b; Alothman et al., 2022). Restoring regular operations and ensuring the impacted systems are safe and operational are the two main goals of recovery. In this phase, it is crucial to perform system integrity tests, monitor restored systems for enduring threats, and restore data using confirmed backups (Carias et al., 2020a, 2020b; Alothman et al., 2022). Organisations that regularly test their backup and recovery procedures can recover from accidents 72% faster than those that do not, per a Ponemon Institute study (Carias et al., 2020a, 2020b; Alothman et al., 2022). After an incident has been resolved, a comprehensive assessment is required to assess the response's efficacy and pinpoint areas that require improvement. This evaluation should examine what went wrong, what worked, and how future occurrences of the same kind might be avoided. To make sure that businesses are better equipped to handle future occurrences, the lessons learned from these evaluations should guide modifications to the IRP, security policies, and training programmes (Carias et al., 2020a, 2020b; Alothman et al., 2022; Chockalingam & Maathuis, 2022). Several established frameworks guide the development and implementation of rapid response and recovery plans, as shown in Figure 10.2.

When creating policies for handling cybersecurity issues, the NIST is essential. A key resource outlining best practices for incident response is the NIST "Computer Security Incident Handling Guide" (SP 800–61), highlighting the significance of a methodical approach to identifying, detecting, preventing, responding to, and recovering from cyber incidents. This manual describes the incident response lifecycle and offers comprehensive suggestions for organising and carrying out successful IRPs, covering topics like reporting, evaluation, and ongoing enhancement of issue-handling procedures (Cichonski et al., 2012). Additionally, enterprises can improve their cybersecurity posture by implementing the NIST Cybersecurity methodology (NIST CSF), a complete methodology. It highlights the necessity of ongoing assessment and enhancement of security procedures and offers an organised method for

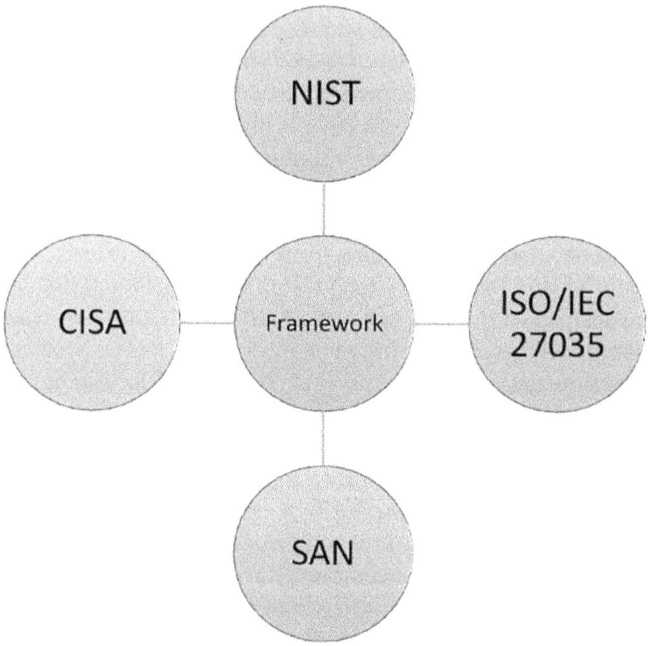

FIGURE 10.2 Framework for response and recovery plans showing how an organisational incident-handling framework can be developed by integrating guidance from major reference models, including NIST, ISO/IEC 27035, CISA, and SAN, to ensure consistent preparation, response, and post-incident improvement.

handling cybersecurity risks (Alraddadi, 2023). NIST recommendations must be included in organisational strategies to reduce the possible financial, operational, and reputational effects of cyber incidents by enabling a methodical response to cyber threats (Nair, 2023).

The SANS Institute has created a framework that describes a six-step incident response procedure and NIST's contributions: preparation, identification, containment, eradication, recovery, and lessons learned. Effective incident management requires careful documentation throughout the incident lifecycle and proactive planning, which are emphasised by this approach (Teh et al., 2021). Businesses that employ these frameworks are better able to react to cyberattacks quickly and efficiently, which reduces disruptions and increases resistance to future threats (Newhouse et al., 2017). The CISA also offers customised playbooks for critical infrastructure sectors, such as water systems and federal agencies. These playbooks provide companies with the tools and tactics they need to successfully handle possible threats by providing practical advice for handling cyber incidents (Stouffer et al., 2015). By utilising these extensive frameworks and recommendations, organisations can significantly improve their incident response skills and protect their vital assets from the constantly changing cyber threat landscape. Investing in strong incident response frameworks is not only a strategic decision but also a business requirement, given the anticipated rise in cybercrime expenses, which are predicted to reach $10.5 trillion annually by 2025 (Calder, 2018). Businesses that put preparedness, detection, containment, eradication, recovery, and post-incident evaluation first are better equipped to lessen the negative consequences of cyberattacks and preserve business continuity in a world that is becoming increasingly digital.

10.9 CYBER RESILIENCE STRATEGIES AND RECOVERY PLANNING FOR WATER INFRASTRUCTURE

Cyber resilience is the ability of an organisation to prepare for, withstand, recover from, and adapt to cyber incidents. For water infrastructure, a critical sector often targeted by cyberattacks, implementing effective cyber resilience strategies and recovery plans is essential to ensure the continuity of operations, protect public health, and maintain trust in water systems. Water utilities increasingly adopt IoT, SCADA systems, and cloud platforms, so they become more vulnerable to cyber threats such as ransomware, data breaches, and operational disruptions.

10.9.1 The Importance of Cyber Resilience in Water Infrastructure

Cyberattacks on water infrastructure can have disastrous results, including interrupted services, environmental harm, and serious health hazards from tainted water supplies. The 2021 attack on the Oldsmar water treatment plant in Florida, where a hacker tried to change chemical levels and directly endangered public safety, is a noteworthy instance that highlights these dangers. This event highlights the weaknesses in vital infrastructure, especially in the water industry, which depends increasingly on networked systems that cybercriminals could use (Clark et al., 2018; Hassanzadeh et al., 2020; You, 2022). According to data from the CISA, more than 60% of water utilities had at least one major cyber event over the previous three years, underscoring the need for strong cyber resilience plans. Given their crucial nature and potential for wide-ranging effects, water systems are frequently viewed as attractive targets, and this figure reflects a more significant trend of rising cyber threats against them (Clark et al., 2018; Hassanzadeh et al., 2020; You, 2022). Incorporating cutting-edge networking technologies into water systems creates vulnerabilities because, although they increase operational efficiency, they also make these systems more vulnerable to cyber breaches (Clark et al., 2018; Hassanzadeh et al., 2020). Furthermore, in reaction to these dangers, the legal environment about cybersecurity in the water industry has changed. According to You (2022), Clark et al. (2018), and Malatji et al. (2020), recent high-profile incidents have prompted calls for more comprehensive legislative actions to strengthen defences against cyber threats. Historically, water and wastewater utilities were primarily left to manage their cybersecurity independently. According to You (2022) and Clark et al. (2018), these legislative actions are essential for creating a framework that encourages the adoption of efficient cybersecurity procedures and improves the general resilience of water infrastructure against possible cyberattacks. Because of the significant effects that cyberattacks have on water infrastructure, urgently improving cyber resilience is necessary. The CISA analysis highlights the sector's pervasive weaknesses, while the Oldsmar incident serves as a sobering reminder of the possible risks. In light of growing cyber threats, legislative measures and strategic investments in cybersecurity are necessary to protect environmental integrity and public health.

10.9.2 Cyber Resilience Strategies for Water Infrastructure

Regular risk assessments are fundamental to a robust cyber resilience strategy, particularly for water utilities that must navigate both cyber and physical vulnerabilities. Identifying critical assets and evaluating their vulnerabilities allows organisations to prioritise risks based on potential impacts, which is essential in hybrid environments where both digital and physical systems are interconnected (Lis & Mendel, 2019). Integrating cyber and physical components in risk assessments is crucial, as many water systems face threats that can exploit weaknesses in either domain (Lis & Mendel, 2019).

A multi-layered security approach is vital for protecting against diverse attack vectors. This includes the implementation of firewalls, IDS, endpoint security, network segmentation, and regular software patching (Ghaib, 2024). For SCADA systems, the integration of security protocols such as MFA and encryption is critical to prevent unauthorised access, ensuring that sensitive operational data remains secure (Paul & Rao, 2022). The Zero Trust Architecture further enhances security by operating on the principle of "never trust, always verify", which is particularly relevant in environments where insider threats may exist (Paul & Rao, 2022; Nkoro, 2024).

Regular and secure backups of critical data and configurations are essential for recovery during a cyber incident. Backup systems should be encrypted, tested, and stored offline or in geographically separate locations to mitigate risks from ransomware attacks or physical disasters (Ghaib, 2024). According to IBM's 2023 "Cost of a Data Breach" report, organisations that have tested backups can significantly reduce recovery times, thereby minimising operational downtime (Ghaib, 2024). This highlights the importance of proactive measures in cybersecurity strategies.

Adopting AI-powered tools for real-time threat detection is another critical component of a comprehensive cyber resilience strategy. These tools can identify anomalies such as unauthorised access or unusual system behaviour, enabling utilities to respond to threats before significant damage occurs (Ghaib, 2024). AI and ML technologies can analyse vast amounts of data to predict potential vulnerabilities, providing a proactive defence against emerging threats (Ghaib, 2024).

A well-developed IRP is crucial for minimising the impact of cyberattacks. The IRP should clearly define roles, responsibilities, communication protocols, and escalation procedures for responding to incidents (Ghaib, 2024). Regular training and simulations, such as tabletop exercises, are necessary to ensure that the plan remains compelling and actionable. Collaboration with government agencies, cybersecurity organisations, and industry peers is also essential for sharing threat intelligence and best practices, with platforms like the WaterISAC facilitating real-time information exchange.

Addressing human error, a leading cause of cyber incidents, is vital. Regular cybersecurity training for employees, including recognising phishing attempts and understanding security protocols, can significantly reduce the risk of accidental breaches. According to Proofpoint's 2023 report, organisations with robust employee training programmes experienced 60% fewer phishing-related incidents. This underscores the importance of fostering a security-conscious organisational culture to enhance cyber resilience.

10.9.3 Recovery Planning for Water Infrastructure

Disaster Recovery Plans (DRPs) are critical frameworks that outline procedures for restoring operations and recovering data following incidents, particularly for essential services such as water utilities. These plans must include restoring SCADA systems, ensuring water quality, and resuming distribution services. Integrating these elements is essential for maintaining operational continuity and safeguarding public health during emergencies (Gudimetla, 2024). Regular testing of the DRP is vital to ensure its effectiveness; organisations that pre-test their IRPs can recover significantly faster—up to 2.5 times quicker—than those without such plans (Sawalha, 2021).

Business Continuity Planning (BCP) complements DRP by ensuring that essential services, including water treatment and distribution, remain operational during disruptions, such as cyberattacks. Effective BCP involves identifying backup facilities, alternate water sources, and manual operations that can temporarily replace automated systems (Sawalha, 2021). The importance of collaboration is underscored by findings from the CISA, which indicate that water utilities with established partnerships and threat-sharing agreements experience fewer successful cyberattacks, highlighting the value of collective resilience efforts (Sawalha, 2021).

The post-recovery analysis is crucial for identifying the root causes of incidents, evaluating response effectiveness, and updating security measures to prevent future occurrences. This reflective process should inform updates to resilience strategies and recovery plans, ensuring that lessons learned contribute to improved practices (Kennedy et al., 2021). Compliance with industry regulations and frameworks such as the NIST Cybersecurity Framework and ISO/IEC 27001 provides a structured approach to recovery and long-term resilience, ensuring accountability and transparency in operations (Gudimetla, 2024).

The financial implications of cyberattacks on critical infrastructure are significant, with the average downtime reported at 22 days and costs averaging $1.85 million per incident, according to the Ponemon Institute's 2023 report (Gudimetla, 2024). This emphasises water utilities' need to adopt comprehensive cyber resilience strategies and recovery planning frameworks. By combining proactive measures—such as risk assessments and AI-driven monitoring—with robust recovery frameworks, utilities can mitigate the impact of cyber incidents and ensure the uninterrupted delivery of essential services (Gudimetla, 2024). The increasing sophistication of cyber threats necessitates a culture of resilience, leveraging advanced technologies, regulatory frameworks, and multi-stakeholder collaboration to protect vital resources (Sawalha, 2021). Figure 10.3 shows the recovery planning for water infrastructure.

FIGURE 10.3 Recovery planning for water infrastructure, illustrating the iterative cycle from pre-incident preparedness and impact assessment, through emergency response and short-term service restoration, to long-term rehabilitation, resilience enhancement, and continuous updating of recovery plans.

10.9.4 Real-World Case Studies of Cybersecurity Breaches in Water Management

Cybersecurity breaches in water management systems have highlighted the vulnerabilities of vital infrastructure, underscoring the dire repercussions for public health and safety. A clear example is the February 2021 incident at the Oldsmar water treatment plant, in which an attacker gained remote access to the SCADA system and tried to dangerously change the quantities of sodium hydroxide. Weak remote desktop protocols and inadequate password management were blamed for this incident; these are typical weaknesses in water systems that call for MFA and more robust access controls (Rautmare, 2011; Stoian et al., 2014). To reduce the dangers associated with such assaults, the incident emphasises the critical necessity for strong cybersecurity safeguards and efficient incident response procedures (Guan et al., 2011). The hazards presented by antiquated SCADA systems were further illustrated by an incident that occurred in 2016 involving the Kemuri Water Company. Hackers altered chemical levels in water treatment processes and obtained access to operational controls by taking advantage of flaws in online applications and legacy systems. Because improper network segmentation made the SCADA systems vulnerable to outside threats, this breach highlighted the urgent need for network segmentation and the modernisation of outdated systems to meet security standards (Mir & Ramachandran, 2019; Palmer et al., 2020). Many SCADA systems were linked to the Internet without proper security measures, according to Verizon's 2016 Data Breach Digest, which emphasised the need for secure remote access protocols (Kim et al., 2014).

Hackers tried to alter water pressure and flow in Israel's water and sewage systems in April 2020 through coordinated cyberattacks. Despite being foiled, these attempts demonstrated the possibility of significant disruptions associated with nation-state actors. To combat advanced cyber threats, the Israel National Cyber Directorate's study stressed the value of international cooperation and real-time surveillance

(Guan et al., 2013). To protect vital services and public safety, this incident and others highlight the urgent need for proactive threat identification and incident response techniques (Foglietta et al., 2019). Water facilities are vulnerable, as demonstrated by the August 2021 ransomware assault on a California water authority. Important operational data was encrypted by the hack, which interfered with administrative tasks and postponed maintenance plans. The incident demonstrated the financial impact of ransomware, as IBM reported an average cost of $4.54 million per incident, even though the utility could restore its systems without paying the ransom (Chochtoula et al., 2022). The need to keep safe, encrypted backups and train staff to spot phishing attempts—a popular entry point for ransomware attacks—is highlighted by this scenario (Barbosa & Pras, 2010).

Furthermore, insider threats and the necessity of secure authentication systems were brought to light in the 2000 Queensland, Australia, event in which a dissatisfied employee tampered with a sewage system (Shahzad, 2014). Significant flaws in credential management were further highlighted in 2021 when it was discovered that a Tampa Bay water utility login credentials were openly available on the dark web. Strong password restrictions and frequent credential audits are essential, as a 2023 Verizon report revealed that a sizable portion of hacking-related breaches involve compromised credentials (Söğüt & Erdem, 2023).

Together, these case studies highlight common flaws in water management systems, such as dependence on antiquated technology, lax access controls, and the ongoing dangers of ransomware and phishing. Since cybercrime is expected to cost the world $10.5 trillion a year by 2025, water utilities must first put resilience, teamwork, and proactive threat mitigation to safeguard public health and vital services (Roclawski, 2022). To overcome these obstacles and improve the overall security posture of critical infrastructure, cooperation via intelligence-sharing platforms such as WaterISAC is essential (Cagalaban & Kim, 2011).

10.10 CHAPTER SUMMARY AND CONCLUSION

This chapter talks about how important cybersecurity is for protecting water systems. It points out the unique risks and problems the water sector faces as everything gets more connected. It examined how cyber threats affect new water management technologies and the weak spots in devices and networks. It shared real stories, like the attack on the Oldsmar water treatment plant and the problem in Maroochy Shire. These examples showed how serious cyber issues can be and why we must prepare for them.

This chapter also discussed ways to make the systems stronger. Using encryption, firewalls, MFA, and splitting networks into separate parts are smart moves. Having good plans for quick responses and recovery is key to cutting down downtime and reducing the impact of cyberattacks. AI can help spot threats early and act fast. Sharing updates on ongoing threats is also crucial. The chapter discussed how important it is to have firm recovery plans. These plans should include assessing risks, adding layers of protection, and backups. Creating a safety-first culture is important, too. This means training employees, doing regular checks, and following safety rules from organisations like NIST and ISO.

In conclusion, water utilities need to have a solid cyber plan. This means combining tech upgrades, following the rules, and working with others. As cyber threats get smarter and more frequent, protecting our water systems is a top priority. By learning from past mistakes and sticking to best practices, we can manage risks better and keep our essential services safe for everyone now and in the future.

REFERENCES

Adelani, F. (2024a). Theoretical frameworks for the role of AI and machine learning in water cybersecurity: Insights from African and U.S. applications. *Computer Science & IT Research Journal*, *5*(3), 681–692. https://doi.org/10.51594/csitrj.v5i3.928

Adelani, F. (2024b). Theoretical perspectives on cybersecurity challenges in critical water infrastructure: Insights from Africa and the United States. *International Journal of Frontiers in Engineering and Technology Research*, *6*(2), 001–007. https://doi.org/10.53294/ijfetr.2024.6.2.0029

Alothman, B., Alhajraf, A., Alajmi, R., Farraj, R., Alshareef, N., & Khan, M. (2022). Developing a cyber incident exercises model to educate security teams. *Electronics*, *11*(10), 1575. https://doi.org/10.3390/electronics11101575

Alraddadi, A. (2023). Developing an abstraction framework for managing and controlling Saudi banks' cybersecurity threats based on the NIST cybersecurity framework and ISO/IEC 27001. *Journal of Software Engineering and Applications*, *16*(12), 695–713. https://doi.org/10.4236/jsea.2023.1612036

Arkan A. Ghaib (2024). Zenith armor: Advancing security with zero trust measures. *International Journal of Scientific Research in Engineering and Management*, *08*(04), 1–5. https://doi.org/10.55041/ijsrem31326

Barbosa, R., & Pras, A. (2010). Intrusion detection in SCADA networks, 163–166. https://doi.org/10.1007/978-3-642-13986-4_23

Cagalaban, G., & Kim, S. (2011). Towards improving SCADA control systems security with vulnerability analysis, 27–32. https://doi.org/10.1007/978-3-642-22706-6_5

Caithamlová, M. (2024). Operation of water supply systems in the Czech Republic—Risk analysis. *Applied Sciences*, *14*(4), 1572. https://doi.org/10.3390/app14041572

Calder, A. (2018). Nist cybersecurity framework. https://doi.org/10.2307/j.ctv4cbhfx

Camacho, N. (2024). The role of AI in cybersecurity: Addressing threats in the digital age. *JAIGS*, *3*(1), 143–154. https://doi.org/10.60087/jaigs.v3i1.75

Carias, J., Arrizabalaga, S., Labaka, L., & Hernantes, J. (2020a). Cyber resilience progression model. *Applied Sciences*, *10*(21), 7393. https://doi.org/10.3390/app10217393

Carias, J., Borges, M., Labaka, L., Arrizabalaga, S., & Hernantes, J. (2020b). Systematic approach to cyber resilience operationalization in SMEs. *IEEE Access*, *8*, 174200–174221. https://doi.org/10.1109/access.2020.3026063

Chochtoula, D., Ilias, A., Stamatiou, Y., & Makris, C. (2022). Integrating elliptic curve cryptography with the modbus TCP SCADA communication protocol. *Future Internet*, *14*(8), 232. https://doi.org/10.3390/fi14080232

Chockalingam, S., & Maathuis, C. (2022). Ontology for effective security incident management. *International Conference on Cyber Warfare and Security*, *17*(1), 26–35. https://doi.org/10.34190/iccws.17.1.6

Cichonski, P., Millar, T., Grance, T., & Scarfone, K. (2012). Computer security incident handling guide: recommendations of the national institute of standards and technology. https://doi.org/10.6028/nist.sp.800-61r2

Clark, R., Hakim, S., & Panguluri, S. (2018). Protecting water and wastewater utilities from cyber-physical threats. *Water and Environment Journal*, *32*(3), 384–391. https://doi.org/10.1111/wej.12340

Clark, R., Panguluri, S., Nelson, T., & Wyman, R. (2017). Protecting drinking water utilities from cyber threats. *Opflow*, *109*(2), 50–58. https://doi.org/10.5942/jawwa.2017.109.0021

Derhab, A., Gabsi, M., Gumaei, A., Μαγλαράς, Λ., Ferrag, M., Mukherjee, M., … & Khan, F. (2019). Blockchain and random subspace learning-based IDs for SDN-enabled industrial IoT security. *Sensors*, *19*(14), 3119. https://doi.org/10.3390/s19143119

Falch, M., Olesen, H., Skouby, K., Tadayoni, R., & Williams, I. (2023). Cybersecurity strategies for SMEs in the nordic baltic region. *Journal of Cyber Security and Mobility*. https://doi.org/10.13052/jcsm2245-1439.1161

Familoni, B. (2024). Cybersecurity in the financial sector: A comparative analysis of the USA and Nigeria. *Computer Science & IT Research Journal*, *5*(4), 850–877. https://doi.org/10.51594/csitrj.v5i4.1046

Foglietta, C., Masucci, D., Palazzo, C., Santini, R., Panzieri, S., Rosa, L., … & Lev, L. (2019). From detecting cyber-attacks to mitigating risk within a hybrid environment. *IEEE Systems Journal*, *13*(1), 424–435. https://doi.org/10.1109/jsyst.2018.2824252

Goel, R., Kumar, A., & Haddow, J. (2020). Prism: A strategic decision framework for cybersecurity risk assessment. *Information and Computer Security*, *28*(4), 591–625. https://doi.org/10.1108/ics-11-2018-0131

Gordon, L., Loeb, M., Lucyshyn, W., & Zhou, L. (2018). Empirical evidence on the determinants of cybersecurity investments in private sector firms. *Journal of Information Security*, *9*(2), 133–153. https://doi.org/10.4236/jis.2018.92010

Guan, J., Graham, J., & Hieb, J. (2011). A digraph model for risk identification and management in SCADA systems, 150–155. https://doi.org/10.1109/isi.2011.5983990

Guan, J., Graham, J., & Hieb, J. (2013). A digraph model for risk identification and management in SCADA systems, 69–83. https://doi.org/10.1016/b978-0-12-404702-0.00004-5

Gudimetla, S. (2024). Cybersecurity considerations in disaster recovery planning. *International Journal for Research in Applied Science and Engineering Technology*, *12*(5), 3143–3149. https://doi.org/10.22214/ijraset.2024.62253

Hassanzadeh, A., Rasekh, A., Galelli, S., Aghashahi, M., Taormina, R., Ostfeld, A., … & Banks, K. (2020). A review of cybersecurity incidents in the water sector. *Journal of Environmental Engineering*, *146*(5). https://doi.org/10.1061/(ASCE)EE.1943-7870.0001686

Kanthimathinathan, A. (2023). A novel cyber resilience framework – Strategies and best practices for today's organizations. *International Journal on Recent and Innovation Trends in Computing and Communication*, *11*(8s), 86–96. https://doi.org/10.17762/ijritcc.v11i8s.7178

Karunamurthy, A. (2023). Human-in-the-loop intelligence: Advancing AI-centric cybersecurity for the future. *Quing International Journal of Multidisciplinary Scientific Research and Development*, *2*(3), 20–43.

Kennedy, M., Gonick, S., & Errett, N. (2021). Are we ready to build back "healthier?" an exploratory analysis of U.S. state-level disaster recovery plans. *International Journal of Environmental Research and Public Health*, *18*(15), 8003. https://doi.org/10.3390/ijerph18158003

Kim, S., Cho, D., & Yeo, S. (2014). Secure model against APT in m-connected SCADA network. *International Journal of Distributed Sensor Networks*, *10*(6), 594652. https://doi.org/10.1155/2014/594652

Kumar, N., Sen, A. C., Hordiichuk, V., Jaramillo, M. T. E., Molodetskyi, B., & Kasture, A. B. (2023). AI in cybersecurity: Threat detection and response with machine learning. *TJJPT*, *44*(3), 38–46. https://doi.org/10.52783/tjjpt.v44.i3.237

Labu, M. (2024). Next-generation cyber threat detection and mitigation strategies: A focus on artificial intelligence and machine learning. *Journal of Computer Science and Technology Studies*, *6*(1), 179–188. https://doi.org/10.32996/jcsts.2024.6.1.19

Lis, P., & Mendel, J. (2019). Cyberattacks on critical infrastructure: An economic perspective. *Economics and Business Review*, *5*(2), 24–47. https://doi.org/10.18559/ebr.2019.2.2

Malatji, M., Marnewick, A. L., & von Solms, S. (2020). Cybersecurity policy and the legislative context of the water and wastewater sector in South Africa. *Sustainability*, *13*(1), 291. https://doi.org/10.3390/su13010291

Malatji, M., & Solms, S. (2020). Cybersecurity policy and the legislative context of the water and wastewater sector in South Africa. *Sustainability*, *13*(1), 291. https://doi.org/10.3390/su13010291

Malek, A. (2021). Bigger is always not better, less is more, sometimes: The concept of data minimization in the context of big data. https://doi.org/10.31124/advance.14601510.v1

Mir, A., & Ramachandran, R. (2019). Security gaps assessment of smart grid based SCADA systems. *Information and Computer Security*, *27*(3), 434–452. https://doi.org/10.1108/ics-12-2018-0146

Mmaduekwe, U. (2024). Bias and fairness issues in artificial intelligence-driven cybersecurity. *Current Journal of Applied Science and Technology*, *43*(6), 109–119. https://doi.org/10.9734/cjast/2024/v43i64391

Moraitis, G., Sakki, G., Karavokiros, G., Nikolopoulos, D., Tsoukalas, I., Kossieris, P., … & Makropoulos, C. (2023). Exploring the cyber-physical threat landscape of water systems: A socio-technical modelling approach. *Water*, *15*(9), 1687. https://doi.org/10.3390/w15091687

Naidoo, R., & Jacobs, C. (2023). Cyber warfare and cyber terrorism threats targeting critical infrastructure: A HCPS-based threat modelling intelligence framework. *European Conference on Cyber Warfare and Security*, *22*(1), 311–318. https://doi.org/10.34190/eccws.22.1.1443

Nair, P. (2023). Enhancing cybersecurity awareness training through the NIST framework. *IJARCCE*, *12*(12). https://doi.org/10.17148/ijarcce.2023.121203

Newhouse, W., Keith, S., Scribner, B., & Witte, G. (2017). National initiative for cybersecurity education (nice) cybersecurity workforce framework. https://doi.org/10.6028/nist.sp.800-181

Nikolopoulos, D., Moraitis, G., & Makropoulos, C. (2021). 7. Strategic and tactical cyber-physical security for critical water infrastructures. https://doi.org/10.1561/9781680838237.ch7

Nkoro, E. (2024). Zero-trust marine cyberdefense for IoT-based communications: An explainable approach. *Electronics*, *13*(2), 276. https://doi.org/10.3390/electronics13020276

Obi, O. (2024). Comprehensive review on cybersecurity: Modern threats and advanced defense strategies. *Computer Science & IT Research Journal*, *5*(2), 293–310. https://doi.org/10.51594/csitrj.v5i2.758

Onih, V. (2024). The role of AI in enhancing threat detection and response in cybersecurity infrastructures. *International Journal of Scientific and Management Research*, *07*(04), 64–96. https://doi.org/10.37502/ijsmr.2024.7404

Onwubiko, C., & Ouazzane, K. (2019). Challenges towards building an effective cyber security operations centre. *International Journal on Cyber Situational Awareness*, *4*(1), 11–39. https://doi.org/10.22619/ijcsa.2019.100124

Palmer, I., Rogers, E., & Mcfly, S. (2020). A graph-based analysis of industrial control systems network traffic. *IJICSS*, *3*(1), 100–108. https://doi.org/10.20533/ijicss.9781.9083.20346.2020.0011

Paul, B., & Rao, M. (2022). Zero-trust model for smart manufacturing industry. *Applied Sciences*, *13*(1), 221. https://doi.org/10.3390/app13010221

Pliatsios, D., Sarigiannidis, P., Λάγκας, Θ., & Sarigiannidis, A. (2020). A survey on SCADA systems: Secure protocols, incidents, threats and tactics. *IEEE Communications Surveys & Tutorials*, *22*(3), 1942–1976. https://doi.org/10.1109/comst.2020.2987688

Rangaraju, S. (2023). AI sentry: Reinventing cybersecurity through intelligent threat detection. *EPH - International Journal of Science and Engineering*, *9*(3), 30–35. https://doi.org/10.53555/ ephijse.v9i3.211

Rantos, K., Spyros, A., Παπανικολάου, A., Kritsas, A., Ilioudis, C., & Katos, V. (2020). Interoperability challenges in the cybersecurity information sharing ecosystem. *Computers*, *9*(1), 18. https://doi.org/10.3390/computers9010018

Rautmare, S. (2011). Scada system security: Challenges and recommendations. https://doi.org/10.1109/indcon.2011.6139567

Rizvi, M. (2023). Enhancing cybersecurity: The power of artificial intelligence in threat detection and prevention. *International Journal of Advanced Engineering Research and Science*, *10*(5), 055–060. https://doi.org/10.22161/ijaers.105.8

Roclawski, H. (2022). Iot.h2o – Supervision and control for water systems based on low-cost hardware and open-source software. https://doi.org/10.4995/wdsa-ccwi2022.2022.14788

Sawalha, I. (2021). Views on business continuity and disaster recovery. *International Journal of Emergency Services*, *10*(3), 351–365. https://doi.org/10.1108/ijes-12-2020-0074

Seker, E., & Ozbenli, H. (2018). The concept of cyber defence exercises (CDX): Planning, execution, evaluation, 1–9. https://doi.org/10.1109/cybersecpods.2018.8560673

Shahana, A. (2024). AI-driven cybersecurity: Balancing advancements and safeguards. *Journal of Computer Science and Technology Studies*, *6*(2), 76–85. https://doi.org/10.32996/jcsts.2024.6.2.9

Shahzad, N. (2014). A new cloud based supervisory control and data acquisition implementation to enhance the level of security using testbed. *Journal of Computer Science*, *10*(4), 652–659. https://doi.org/10.3844/jcssp.2014.652.659

Shoetan, P. (2024). Synthesizing AI's impact on cybersecurity in telecommunications: A conceptual framework. *Computer Science & IT Research Journal*, *5*(3), 594–605. https://doi.org/10.51594/csitrj.v5i3.908

Sinha, A. (2023). Artificial intelligence and machine learning for cybersecurity applications and challenges, 109–146. https://doi.org/10.4018/978-1-6684-9317-5.ch007

Söğüt, E., & Erdem, O. (2023). A multi-model proposal for classification and detection of ddos attacks on SCADA systems. *Applied Sciences*, *13*(10), 5993. https://doi.org/10.3390/app13105993

Stoian, I., Ignat, S., Capatina, D., & Ghiran, O. (2014). Security and intrusion detection on critical SCADA systems for water management. https://doi.org/10.1109/aqtr.2014.6857919

Stouffer, K., Falco, J., & Scarfone, K. (2015). Guide to industrial control systems (ICS) security. https://doi.org/10.6028/nist.sp.800-82r2

Taherdoost, H. (2022). Understanding cybersecurity frameworks and information security standards—A review and comprehensive overview. *Electronics*, *11*(14), 2181. https://doi.org/10.3390/electronics11142181

Teh, Y., Tan, Y., & Wong, S. (2021). 5G cybersecurity: Risk assessment and incident response in the healthcare industry, 145–152. https://doi.org/10.56453/icdxa.2021.1015

Temara, S. (2024). Harnessing the power of artificial intelligence to enhance next-generation cybersecurity. https://doi.org/10.36227/techrxiv.170785703.32137017/v1

Tuptuk, N., Hazell, P., Watson, J., & Hailes, S. (2021). A systematic review of the state of cyber-security in water systems. *Water*, *13*(1), 81. https://doi.org/10.3390/w13010081

Vielberth, M., Böhm, F., Fichtinger, I., & Pernul, G. (2020). Security operations center: A systematic study and open challenges. *IEEE Access*, *8*, 227756–227779. https://doi.org/10.1109/access.2020.3045514

Wai, E. (2023). Seamless industry 4.0 integration: A multilayered cyber-security framework for resilient SCADA deployments in CPPS. *Applied Sciences*, *13*(21), 12008. https://doi.org/10.3390/app132112008

Wali, A. (2024). A survey of security challenges in cloud-based SCADA systems. *Computers*, *13*(4), 97. https://doi.org/10.3390/computers13040097

You, J. (2022). Strengthening cybersecurity of water infrastructure through legislative actions. *JAWRA: Journal of the American Water Resources Association*, *58*(2), 282–288. https://doi.org/10.1111/1752-1688.12995

Zakroum, M., Jérôme, F., Chrisment, I., & Ghogho, M. (2022). Monitoring network telescopes and inferring anomalous traffic through the prediction of probing rates. *IEEE Transactions on Network and Service Management*, *19*(4), 5170–5182. https://doi.org/10.1109/tnsm.2022.3183497

Zohuri, B. (2024). AI revolution: Safeguarding tomorrow's frontiers – Transforming cybersecurity across industries (a short approach). *Current Trends in Engineering Science*, *4*(2), 1–4. https://doi.org/10.54026/ctes/1057

11 Water 4.0 *Opportunities and Challenges*

11.1 INTRODUCTION

Water 4.0 refers to integrating advanced technologies and digital automation into water management systems to optimise water distribution and improve efficiency. This approach, also known as Intelligent Water Networks, presents several opportunities for the water industry (Joseph et al., 2022). Water 4.0 presents an opportunity to enhance water management systems' efficiency, sustainability, and resilience, leading to improved service delivery, reduced costs, and better environmental outcomes (Alabi et al., 2019). Using information science to improve environmental science and technology is a hot topic in the continuing "big data" discussion. The Water 4.0 framework is a novel conference subject because it integrates environmental studies and information technology to improve cyber-physical systems used in water management.

Making the most of environmental data in the Environmental Information System (EIS) requires the creation of specialised procedures. Scientific visualisation allows for the easy amalgamation of enormous volumes of disparate environmental data set against a realistic geographical backdrop. Information science and technology like visual data analytics, machine learning, and artificial intelligence (AI) may aid future developments in the sector. Together, environmental science and information technology (through EISs and Water 4.0 ideas) provide an exciting opportunity for collaborative study (Adedeji et al., 2022).

11.2 OPPORTUNITIES OF WATER 4.0

Szádeczky (2023) addresses the growing cyber threat posed by smart water management in the context of the Fourth Industrial Revolution, focusing on Hungary and Central Europe. According to him, Water 4.0, also known as Smart Water or Digital Water, offers several practical advantages and opportunities for water management. It aims to transform municipal water management's infrastructure and operation by incorporating advanced digital technologies and data-driven approaches. Some of the opportunities associated with Water 4.0 include:

Economic Efficiency: Water 4.0 solutions can lead to lower operating costs (OC), higher revenues, and better return on investment, resulting in increased economic value for water utilities. These solutions optimise processes, enhance customer satisfaction, and enable more economically efficient operations.

DOI: 10.1201/9781003633839-11

Environmental Protection: Water 4.0 solutions contribute to environmental sustainability by improving the efficiency and sustainability of water services and ecosystems. By optimising the planning, operation, and maintenance of water resources and infrastructure, these solutions help mitigate water scarcity, pollution, climate change, and ageing infrastructure challenges.

Process Optimisation: Integrating digital technologies in water management enables predictive maintenance, efficient resource allocation, and process optimisation. By leveraging data-driven approaches, water utilities can increase operational security, achieve higher levels of availability, and implement more effective human resource management.

Regulatory Compliance: Water 4.0 solutions facilitate regulatory compliance by providing real-time monitoring, analysis, and reporting capabilities. This helps water utilities meet legal requirements and quality standards more effectively.

Other related opportunities in Water 4.0, according to Szádeczky (2023) include:

1. Optimisation of raw water intake by measurement.
2. Low pumping energy consumption.
3. Monitoring the quality of reservoir water.
4. Estimating the volume and quality of raw water to be processed.
5. Chemical, chlorine gas, and alumina dosage optimisation in water purification systems.

Furthermore, Adedeji et al. (2022) explain that the Water 4.0 framework offers several opportunities in water management. Its primary goal is the automated operational management of water systems, through which existing infrastructures may be better managed and optimised. Modern ideas from information science and technology, such as visual data analytics, machine learning methodologies, and AI, may be used to develop digitally integrated water management according to the Water 4.0 concept. According to him, some specific opportunities offered by Water 4.0 include:

Improved Efficiency: Water 4.0 can help increase the efficiency of water systems by automating the management processes and optimising the utilisation of existing infrastructure.

Real-Time Monitoring: Water 4.0 enables real-time monitoring of water systems using sensors and remote-sensing technologies. This allows for continuous data collection, which can be integrated into a virtual water system for analysis and decision-making.

Predictive Capabilities: The Water 4.0 framework emphasises the ability to predict changes within the environment using continuous data streams for model validation. By incorporating modelling tools, Water 4.0 enables the prediction of hydrological environments and their behaviour.

Integration of Big Data: Water 4.0 leverages the ongoing big data debate and utilises information science to handle large amounts of heterogeneous environmental data. This integration enables the extraction of valuable insights and the improvement of data value through machine learning and AI.

Collaboration and Cooperation: Water 4.0 provides a new research field for cooperation between environmental research and information technology. It fosters collaboration between practitioners, stakeholders, policymakers, and researchers to develop and advance the digitally mediated water management concept.

Also, Alabi et al. (2019) in their work explained that Water 4.0, also known as Smart Water or Digital Water, offers several opportunities for the water industry. Some of these opportunities include:

Resource-Efficient Water Management: Water 4.0 utilises digitalisation and automation to optimise the use of water resources. By implementing smart technologies, such as the Internet of Things (IoT), water consumption can be better monitored and managed in real time, leading to more efficient water usage.

Improved Water Security: Water 4.0 enables greater transparency and real-time water supply and demand data access. This information can be used to ensure the availability and reliability of water resources, making the water supply more secure.

Increased Service Quality and User Satisfaction: Water utilities can identify patterns and trends in water usage by leveraging technologies like big data analytics and AI. This enables proactive maintenance and repair, enhancing service quality and, ultimately, higher customer satisfaction.

Environmental Awareness and Sustainability: Water 4.0 promotes public awareness about the environment and the water cycle. Through digitalisation, individuals can better understand their water consumption patterns and the impact of their actions on the environment. This knowledge can drive sustainable behaviours and conservation efforts.

Transfer of Knowledge and Improved Corporate Image: The adoption of Water 4.0 technologies helps water industry players position themselves as progressive and environmentally conscious organisations. It provides an opportunity for knowledge sharing and collaboration among stakeholders, leading to innovation and improved reputation.

According to Joseph et al. (2022), one opportunity is the implementation of digital metering solutions. These meters enable real-time monitoring of water usage, allowing for better management of resources and early detection of leaks or anomalies in the system. Utilities can implement targeted conservation measures and reduce wastage by identifying areas of high water consumption. Water quality and pressure sensors are another opportunity in Water 4.0. These sensors monitor water quality parameters and pressure levels in the distribution network. This data can detect contamination events or pressure variations that could affect service delivery. By acting proactively and promptly addressing these issues, water utilities can ensure the supply of clean and safe water to consumers. Using big data systems and decision support systems is another significant opportunity in Water 4.0. These technologies enable vast amounts of data from various sources, such as sensors, meters, and customer feedback. By utilising advanced analytics and machine learning algorithms, water utilities can gain valuable insights into system operations, optimise resource allocation, and make informed decisions regarding maintenance and infrastructure upgrades.

Asset optimisation is yet another opportunity in the context of Water 4.0. By utilising real-time data and predictive analytics, water utilities can optimise the lifespan and performance of their assets, such as pumps, pipes, and treatment plants. This proactive approach helps identify potential failures or maintenance needs in advance, improving reliability and reducing operational costs.

According to Grooms (2018), in his review of the book *Water 4.0: The Past, Present, and Future of the World's Most Vital Resource* by David Sedlak, several opportunities are associated with Water 4.0. One of the opportunities mentioned is re-examining innovations from ancient Rome's water systems for use in modern urban water projects. Another opportunity is exploring new ideas for drinking water, such as the "toilet to tap" concept and desalination. Furthermore, Sedlak discussed the potential of policy choices resulting in lower water usage and technologies allowing water recycling. These opportunities aim to address the challenges faced by urban water systems and ensure the sustainable provision of clean water in the future (Sedlak, 2014).

Duan et al. (2015) discussed the concept of green chemistry for nanoparticle synthesis, emphasising the use of environmentally benign solvents. Water is highlighted as a favourable solvent choice in green chemistry due to its non-toxic and non-flammable nature. The use of water as a solvent is economically advantageous and environmentally favourable. It is also mentioned that the best solvent is no solvent, but water is preferred if a solvent is needed. Water can be used as a solvent in the synthesis of nanoparticles, mainly through supercritical hydrothermal synthesis. This method utilises water in its supercritical state, allowing for the dissolution of nonpolar organics and the subsequent synthesis of nanoparticles. The dielectric constant of water in the supercritical state plays a crucial role in controlling the formation of nanoparticles. Further advancements in water-based solvents have been explored, including using supercritical carbon dioxide (scCO2) as a solvent alternative. Like water, scCO2 is non-toxic, non-flammable, and environmentally benign. It also offers energy efficiency and easy recovery and reuse.

Overall, these opportunities arise from integrating Industry 4.0 technologies, such as IoT, big data and analytics, cloud computing, and AI, into the water industry. Water 4.0 aims to revolutionise the sector, making it more efficient, sustainable, and responsive to the needs of the growing population and changing environmental conditions. The use of water as a solvent, particularly in its supercritical state, presents opportunities for green chemistry and the synthesis of nanoparticles. Adopting water-based solvents, along with other environmentally benign alternatives, can contribute to developing sustainable and energy-efficient synthesis methods. These opportunities highlight the potential for Water 4.0 to revolutionise water management by integrating environmental sciences and information technology. Moreover, they may aid in the development of essential water infrastructure in a way that is both secure and long-term.

11.3 CHALLENGES OF WATER 4.0

Water 4.0/Smart water management posed some challenges (Szádeczky, 2023). These challenges include various barriers and obstacles to its broader adoption and impact. According to Szádeczky (2023), challenges include:

Lack of Awareness, Knowledge, and Skills: Decision-makers, stakeholders, and users often lack awareness and understanding of smart water technologies and their benefits.

High Costs, Risks, and Uncertainties: Investing in smart water projects and assets can be expensive and comes with risks and uncertainties, which can hinder their implementation.

Limited Availability and Accessibility of Data: The availability and accessibility of data, information, and tools for water management decision-making and performance evaluation are sometimes limited, affecting the effectiveness of Water 4.0 initiatives.

Fragmentation and Heterogeneity of Water Governance and Management: The fragmentation and heterogeneity of water governance and management structures, processes, and systems pose a challenge to implementing and coordinating smart water management approaches.

Lack of Collaboration and Coordination: There is a need for more collaboration, coordination, and communication between various actors and sectors involved in smart water management, including governments, utilities, industries, and research institutions.

According to Adedeji et al. (2022), the challenges of Water 4.0 include:

Automatic Operational Management: When controlling and optimising already-in-place infrastructure, Water 4.0 is primarily concerned with the autonomous operational management of water systems. However, this idea has only just begun to be put into practice; therefore, further development and real-world case studies are needed to establish its efficacy.

Involvement of Practitioners, Stakeholders, and Policymakers: While advancements in computer science are essential, the success of Water 4.0 ideas also needs the participation of practitioners, stakeholders, and politicians. Collaboration and cooperation between stakeholders are essential to implement Water 4.0 solutions effectively.

Development of Tailored Workflows: The best possible use of available environmental data is dependent on the technological creation of purpose-built procedures. In a practical geographical setting, these procedures must be crafted to integrate and analyse massive volumes of diverse environmental data.

Integration of Big Data and Information Science: The environmental sciences and technologies are the only areas that have benefited from the increased scrutiny resulting from the continuing considerable data discussion. Developing functional information systems and enhancing data value relies on incorporating big data and related ideas, such as machine learning and AI.

Developing Cyber-Physical Systems: Water 4.0 is based on developing and applying cyber-physical systems. The challenge lies in effectively demonstrating the application of these systems for water management and automation of water infrastructure.

Predictability of Hydrological Environments: The concept of Water 4.0 relies on the predictability of hydrological environments by combining environmental sciences with information technology. This requires the development and integration of established modelling tools to forecast the behaviour of water systems accurately.

Advances in Computer Hardware: New developments in computer hardware need to be considered to utilise the available computational power for more refined and precise process simulations. Keeping up with advancements in hardware, such as exascale computing, is essential for developing Water 4.0.

According to Alabi et al. (2019), the challenges of Water 4.0 include population growth and the need to meet increasing water supply demands in urban areas due to urban sprawl. The water industry is also facing various global challenges. Additionally, transitioning to digital water management systems and implementing Industry 4.0 technologies requires careful planning and change management in the water industry. Ensuring the security of digital systems from cyberattacks is another significant challenge. Furthermore, integrating data-driven technologies like the IoT, big data and analytics, cloud computing, and AI into the water industry requires investment in infrastructure and training. The complexity of the water industry and the need for an integrated business model add to the challenges of Water 4.0.

In their work, Joseph et al. (2022) stated that Water 4.0 refers to the concept of utilising advanced technologies and digital automation to optimise water management and ensure long-term sustainability. The challenges faced by Water 4.0 include:

Ageing Infrastructure: Traditional water supply systems often suffer from deteriorating water pipe networks, resulting in water losses through leakages and bursts. The average rate of system rehabilitation and upgrading does not keep up with expanding needs, quality demands, and ageing infrastructures. Upgrading and modernising the infrastructure is crucial for implementing Water 4.0.

Water Scarcity: Rapid population growth and urbanisation contribute to the depletion of water resources. Climate change and droughts further exacerbate the challenge of water scarcity. Water management systems need to be more sustainable and efficient to meet the increasing water demands.

Water Losses: Quantifying and addressing water losses through leakages is a significant challenge in water management. Traditional systems often face difficulties in identifying the location of leakages and responding to them promptly. Water 4.0 systems aim to leverage real-time monitoring and analytics to detect and minimise water losses.

Environmental Effects: The water industry faces the challenge of adopting management systems with minimal environmental impact, including reducing greenhouse gas emissions. Adopting water management practices that are more environmentally friendly, including reducing energy consumption and optimising pumping operations, is crucial for Water 4.0.

Grooms (2018) affirmed that there are several challenges associated with the Water 4.0 system. The financial investment required to filter and distribute potable water adequately is one such obstacle. Utilities, according to him, need to invest heavily in updating their vital infrastructure, which comes with several expenditures, including those associated with the workforce, delayed pipe maintenance, replacement infrastructure, and energy. Sedlak further notes that changes in water availability due to climate change contribute to the difficulties already encountered by water companies. The book also highlights the increasing difficulty posed by human hormones in wastewater. Some fish species undergo noticeable sex changes because of these hormones, and these hormones may have consequences for other animals and humans as well. While modern methods can eliminate most of these chemicals, problems with chlorine treatment may lead to the formation of additional harmful substances. Activated carbon and ozone are two other treatment strategies that might be used to neutralise this danger. Furthermore, Sedlak emphasises the need for policy choices that result in lower water usage and technologies that allow for water recycling. This necessitates a total rethinking of drinking water provision and the exploration of new ideas such as "toilet to tap" and desalination. However, the implementation of these new technologies and policies may face resistance and require significant changes in current practices and infrastructure. To summarise, the challenges associated with Water 4.0 include economic costs, water availability shifts due to climate change, the presence of human hormones in wastewater, potential health risks from water treatment chemicals, and the need for policy changes to promote water conservation and recycling.

Duan et al. (2015), however, stated that there are challenges associated with its use in nanoparticle synthesis. One challenge is the high heat capacity of water, which requires intensive energy inputs for production, making energy-saving production a challenge. Additionally, the dielectric constant of water impacts the reaction rate, equilibrium, and solubility, which in turn controls the formation of nanoparticles. Another challenge is corrosion, which can hinder the commercial utilisation of supercritical hydrothermal synthesis using water as a solvent.

These challenges, although significant, can be addressed with proper awareness, investment, data accessibility, and collaboration between stakeholders to achieve the full potential of Water 4.0 for improved efficiency, sustainability, and resilience of water services and ecosystems. It is important to address these challenges to successfully implement Water 4.0 and create digitally mediated water management systems.

11.4 PROGRESS AND FUTURE DIRECTIONS

Water 4.0, also known as smart water management, is a digital transformation of water management that utilises advanced digital technologies and data-driven approaches to optimise the planning, operation, and maintenance of water resources and infrastructure. It aims to improve the efficiency, sustainability, and resilience of water services and ecosystems. The progress of Water 4.0 is driven by various factors, including the growing challenges of water scarcity, pollution, climate change, and ageing infrastructure, as well as the increasing expectations of consumers, regulators, and stakeholders for better quality, affordability, and accountability in water management. The practical advantages of Water 4.0 include lower operating costs, better return on investment, higher revenues, and more economically efficient operation of water utilities. It enables environmental protection, process optimisation, predictable maintenance, and regulatory compliance. Water 4.0 also offers operational security, higher levels of availability, and more predictable human resource management.

Despite its potential, the broader adoption and impact of Water 4.0 face various barriers and challenges. These include the lack of awareness, knowledge, and skills about smart water technologies and benefits among decision-makers, stakeholders, and users; high costs, risks, and uncertainties associated with investing in smart water projects and assets; limited availability and accessibility of data, information, and tools for water management decision-making and performance evaluation;

fragmentation, heterogeneity, and compatibility issues in water governance and management structures, processes, and systems; and the need for more collaboration, coordination, and communication between various actors and sectors involved in smart water management. To enhance the progress and future direction of Water 4.0, there is a need for increased research, development, and innovation. Future studies should focus on understanding and managing the emerging cybersecurity challenges and opportunities in smart water management, particularly in the context of the Fourth Industrial Revolution and associated cybersecurity risks. Strengthening the security of critical water infrastructure in Hungary and Central Europe requires comprehensive research, enhanced security measures, and improved collaboration between researchers, practitioners, and policymakers to ensure the safe and sustainable development of smart water management.

According to Adedeji et al. (2022), Water 4.0 focuses on automatically managing water systems to control and optimise existing infrastructures. Its development is still in its early stages, and practical case studies are crucial for further advancement and proving the concept. The success of Water 4.0 depends on progress in computer sciences and the involvement of practitioners, stakeholders, and policymakers. The concept of Water 4.0 is closely related to the development of EISs. EISs incorporate modern remote sensing and monitoring techniques beyond traditional Geographic Information Systems (GISs). They include continuous data streams for model validation and the ability to predict changes within the environment. The technical development of tailored workflows for optimal usability of available environmental data is essential.

Future directions for Water 4.0 involve the integration of environmental sciences with information technology, mainly through EIS and Water 4.0 concepts. This collaboration between environmental research and information science will further pave the way for digitally mediated water management. Moreover, new developments in computer hardware, such as increased computational power, will enable more refined and precise process simulations. This will require considering both development- and application-oriented principles.

Alabi et al. (2019) stated that the progress and future direction of Water 4.0 involves the digital transformation of the water industry using Industry 4.0 technologies. The goal of Water 4.0 is to make water management systems more efficient, less wasteful, and more sustainable. The water sector is struggling now for various reasons, including the rising global population. Considering these challenges, the water industry is moving towards digital water management systems. The idea behind this shift is Water 4.0, often known as Smart Water. It entails using advanced technologies from the Fourth Industrial Revolution (IoT, Big Data, Cloud, Cybersecurity, and AI). Water 4.0 aims to revolutionise the water sector by improving efficiency and introducing new technologies. The goal is to better use existing data and decentralise water management systems to make better decisions. Water 4.0 has several positive outcomes, including better water and energy management, greater water security, more environmental consciousness among the general population, higher quality services, and information sharing. Moreover, it opens new possibilities for responsible water management in the 21st century.

Water 4.0's long-term goal is to create an industry-wide Integrated Business Model. The technology, business systems, and business strategy dimensions are all considered in this approach. Its goal is to serve as a guide for the water sector while it undergoes its digital revolution. From the perspective of Industry 4.0, there is no integrated business model for Water 4.0. However, it is widely agreed that Water 4.0 cannot succeed without incorporating a wide range of Industry 4.0 technology. In summary, Water 4.0 is facilitating the digital revolution of the water sector. Water 4.0, which uses Industry 4.0 technology, will be implemented by creating an Integrated Business Model. Grooms (2018) asserted in his review of the book *Water 4.0: The Past, Present, and Future of the World's Most Vital Resource* (Sedlak, 2014) that the concept of Water 4.0 involves reimagining and rethinking the provision of drinking water. Sedlak argues that cities must move away from the traditional approach of simply treating drinking water and instead focus on lower water usage, water recycling, and the adoption of new technologies for water provision. He explores ideas such as "toilet to tap" (treating and reusing wastewater for drinking water) and desalination as potential solutions for future water challenges.

Sedlak acknowledges that economic challenges are associated with maintaining and improving water infrastructure. Utilities require substantial investments and updates to their infrastructure, and the costs of labour, pipe maintenance, replacement infrastructure, and energy are high. He also highlights the constraint of water availability shifts due to climate change. As a result, he suggests that water bills may increase as utilities struggle to meet these financial demands. In terms of future technologies, Sedlak discusses the potential of water recycling and the use of alternative treatment technologies such as activated carbon or ozone to combat emerging threats like high concentrations of human hormones in wastewater. He also mentions the need to reconsider the current practice of releasing treated sewage into drinking water sources.

Overall, the future direction of Water 4.0, as outlined by Sedlak, involves a comprehensive rethinking of water provision, with a focus on lower water usage, water recycling, and the adoption of new technologies. It also requires addressing economic challenges and exploring alternative treatment methods to ensure the safety and sustainability of water resources.

11.5 CHAPTER SUMMARY AND CONCLUSION

In summary, Water 4.0 presents opportunities for the water industry to enhance efficiency, improve service delivery, and ensure long-term sustainability. Through the integration of advanced technologies, real-time monitoring, and data-driven decision-making, water utilities can optimise water distribution, reduce losses, and effectively manage resources in the face of growing challenges. However, despite the opportunities, Water 4.0, or smart water management, faces several challenges, including lack of awareness, high costs, risks, and uncertainties, limited data availability and accessibility, fragmentation and heterogeneity of water governance and management, and lack of collaboration and coordination between various actors and sectors. Challenges include automatic operational management, active involvement of practitioners, stakeholders, and policymakers, development of tailored workflows, integration of big data and information science, development of cyber-physical systems, predictability of hydrological environments, and advancements in computer hardware.

The challenges of Water 4.0 include population growth, urban sprawl, global challenges, transitioning to digital water management systems, and ensuring the security of digital systems from cyberattacks. Additionally, integrating data-driven technologies like the IoT, big data analytics, cloud computing, and AI requires investment in infrastructure and training. The complexity of the water industry and the need for an integrated business model further complicate the challenges of Water 4.0. Addressing these challenges requires the integration of technologies such as digital metering solutions, water quality and pressure sensors, big data systems, decision support systems, and asset optimisation within an intelligent water network. These technologies enable efficient water distribution, precise leak detection, and real-time monitoring of water quality and system performance. By leveraging these technologies, Water 4.0 aims to improve service delivery, reduce operational costs, and increase the reliability of water systems.

The progress and future directions of Water 4.0 involve the implementation of practical case studies, the involvement of stakeholders and policymakers, the development of tailored workflows, a collaboration between environmental sciences and information technology, and advancements in computer hardware and computational capabilities.

REFERENCES

Adedeji, K. B., Ponnle, A. A., Abu-Mahfouz, A. M., & Kurien, A. M. (2022). Towards digitalization of water supply systems for sustainable smart city development—Water 4.0. *Applied Sciences*, *12*(18), 9174.

Alabi, M. O., Telukdarie, A., & Janse van Rensburg, N. J. (2019). Water 4.0: An integrated business model from an industry 4.0 approach. *2019 IEEE International Conference on Industrial Engineering and Engineering Management (IEEM)* (pp. 1576–1580). IEEE. https://doi.org/10.1109/IEEM44572.2019.8978859

Duan, H., Wang, D., & Li, Y. (2015). Green chemistry for nanoparticle synthesis. *Chemical Society Reviews*, *44*(16), 5778–5792.

Grooms, K. K. (2018). *Book Review: "Water 4.0: The Past, Present, and Future of the World's Most Vital Resource"*. World Scientific.

Joseph, K., Sharma, A. K., & Van Staden, R. (2022). Development of an intelligent urban water network system. *Water*, *14*(9), 1320.

Sedlak, D. (2014). *Water 4.0: The Past, Present, and Future of the World? S Most Vital Resource*. Yale University Press.

Szádeczky, T. (2023). Water 4.0 in Hungary: Prospects and cybersecurity concerns. *Acta Polytechnica Hungarica*, *20*(7), 195–212. https://doi.org/10.12700/APH.20.7.2023.7.12

12 Concluding Remarks

This book provides a comprehensive examination of Water 4.0. This concept pertains to integrating cutting-edge technologies into water management systems, including cyber-physical systems (CPS), Internet of Things (IoT), cloud computing, artificial intelligence (AI), and automation. This book examines the possible uses of Water 4.0 in improving the operational efficiency of water supply systems. These applications include pipeline health monitoring, pressure control, water quality monitoring, and leakage detection and monitoring. This book established that future water supply networks may benefit considerably from IoT technology due to improvements in efficiency, sustainability, and reliability. When it comes to optimising control actions and managing interactions between various variables in water supply networks, model predictive control (MPC) is a cutting-edge control method.

This book additionally examines the problems and opportunities associated with Water 4.0, such as sustainable water supply and the development of smart cities. These include concerns regarding security, data quality and uncertainty, energy solutions, and regulatory laws. The high cost of using IoT technologies in water distribution systems and cyberattacks on IoT networks are significant obstacles to IoT deployment in water management, and collective action is necessary to overcome these. Water supply network data may be effectively managed using lossless compression methods like run-length encoding (RLE) and Huffman Coding. CPS may be used to monitor water quality, control energy use, and detect problems in water distribution systems. The management of water distribution systems may benefit from IoT, collaboration, sophisticated control tactics, practical data management approaches, and the usage of CPS. This book seeks to offer a thorough examination and profound comprehension of Water 4.0 to inspire more scholarly investigations in this domain.

12.1 SUMMARY OF THIS BOOK

The IoT is a network of physical objects connected to the Internet that can communicate with each other and other systems. Water quality, flow rate, pressure, and temperature data may be collected in real-time using the IoT, allowing operators to remotely monitor the system, automate jobs, and make educated choices. The EU's Digital Single Market initiative is helping to speed up Europe's adoption of IoT by encouraging the growth of digital technologies in the area (Kagermann et al., 2013). Together, the computational and physical components of a CPS carry out a task, improving the system's reliability and sustainability by cutting down on energy waste and water wastage (Chen et al., 2012; Scholes et al., 2016). With the US government's establishment of the Advanced Manufacturing Cooperation to encourage advanced manufacturing technologies like IoT and CPS, the use of these tools is on the rise across a wide range of industries in North America (Pons et al., 2023; Zhou & Wang, 2023).

DOI: 10.1201/9781003633839-12

The sensor, actuation, network, and application layers are all parts of the IoT architecture's application layer. In IoT applications, smart sensors analyse sensor data to provide insights into the effectiveness and efficiency of the system. Gateways for the IoT are hubs for all connected devices, allowing unified management and access to advanced features, including centralised data collection, filtering, processing, and analytics. Information is transferred between gadgets, gateways, and the cloud via wireless communication networks. An Application Management Centre provides real-time visibility into the performance of IoT applications and serves as a platform for managing and monitoring these programmes (Tim et al., 2023). Despite data management, security, cost, and interoperability constraints, many water utilities have adopted solutions to boost network performance and decrease costs by using IoT in water supply network management.

Water distribution networks may benefit significantly from CPS because of their ability to optimise the performance of individual components, such as sensors, pumps, valves, and control systems. Monitoring the network gathers information on water quality, pressure, flow, and other factors to better control pump and valve functions. By anticipating and diagnosing potential problems, data analytics and machine learning methods can lessen network outages and improve service reliability. CPS can monitor and control water quality, fine-tune treatment systems to maintain high water quality and optimise energy use by regulating pump and valve operation to minimise energy waste while maintaining a constant water supply. Physical, computational, and communication aspects comprise CPSs' three pillars. Computational parts process and analyse data using machine learning and control algorithms, while physical parts contain instruments for monitoring and managing physical processes. Connectivity between hardware and software requires communication components like networks and protocols (Lesch et al., 2023; Wani et al., 2023).

Multiple layers of components and capabilities make up CPS architecture. Sensors in CPS, such as those for water flow, pressure, and flow, collect data on the physical process being controlled. Data from sensors is processed by CPS's actuation system, which then takes immediate action. Communication networks are crucial to providing real-time data exchange and decision-making among system components. CPS has been found to be used in a wide range of contexts, including water distribution networks, transportation networks, energy networks, and healthcare delivery networks. Functions like control, communication, and computing all fall within its purview. By "control", we mean the system's capacity to influence the external environment via actuators and sensors, and by "communication", we mean the system's capacity to exchange information among its many parts. Data collected by sensors and other parts of the system is processed by the computer so that it may make decisions and take actions depending on the state of the physical environment (Wani et al., 2023).

CPS can monitor infrastructure and equipment, predict when repairs or maintenance are needed, estimate future water demand, and report actual water use in real time. It may aid water demand predictions, waste reduction, and emergency preparedness. CPS may promote water conservation and smart metering, encouraging consumers to alter their water use patterns. In conclusion, CPS is an effective method that aids in water management, sustainability, and conservation. Water distribution networks, transportation networks, energy networks, and healthcare infrastructures may all benefit from CPS when used in conjunction with the digital world.

The goal of intelligent computing is to optimise computational activities and provide optimum results via the integration of AI with computer technologies. Improvements in processing power, data analytics, and control systems have revolutionised water supply network management, essential for effective and sustainable water distribution. Operators can foresee future needs, anticipate probable faults, and optimise maintenance schedules using predictive analytics, fault detection algorithms, and decision support systems. Smart water grids optimise water distribution, identify leaks, and give real-time information on water consumption by integrating sensors, actuators, and communication networks. To derive meaning from massive datasets, analysts need reliable data management systems, scalable computing frameworks, and cutting-edge analytics tools to conduct big data analysis. Automated telemetry systems, remote sensing technologies, and human data input are all examples of data-gathering techniques. Data-driven analysis, modelling, and decision-making are made possible through data-integration methods (Maier et al., 2014; Zhang et al., 2021).

By facilitating effective operation, preventative maintenance, and enhanced decision-making, intelligent computing has the potential to transform the administration of water supply networks completely. Developments have assisted the processing and integration of disparate data sources in data management methods, cloud computing, and data integration platforms. Security and reliability have been improved using sophisticated modelling and control methods, encrypted communication protocols, intrusion detection systems, and redundant structures. Interfaces and decision-making tools have been made more accessible and simpler using human-centred design concepts.

Monitoring and controlling flow rates, pressure levels, water quality, and equipment conditions in real time is only possible with automation, making it an essential tool for managing water supply networks. MPC, fuzzy logic control, distributed control systems, and Intelligent Control are all examples of cutting-edge control system technologies that improve system performance, efficiency, and dependability. Robotics, autonomous systems, environmental monitoring, healthcare, and smart cities rely heavily on integrated sensors. Management of water distribution networks, including hydroelectric power plants, has significantly benefited from intelligent computers and automation. These innovations allow continuous tracking and analysis, predictive maintenance, defect detection, power production optimisation, and system interaction with energy management software. Water utilities and decision-makers may enhance water quality, conservation efforts, and system resilience by combining data from several sources (Rai et al., 2012; Kim et al., 2014; Mallik et al., 2021).

Benefits such as increased productivity, better decision-making, and more efficient resource management have led to the widespread adoption of intelligent computing and automation technologies in countries like France, Spain, and Singapore. Improvements in sensor deployment and data gathering may be attributed to developments in sensor technology, data acquisition systems, and communication networks. MPC, adaptive control, and optimisation approaches are only examples of state-of-the-art control algorithms designed to handle complex processes. Due to budget limitations, operational limits, and compatibility concerns with older systems, implementing and integrating sophisticated control systems into pre-existing water infrastructure may be difficult. Improvements in the hardware and software of control systems have made it simpler to integrate with pre-existing infrastructure. Successful implementation relies heavily on human factors and operator training (Alabi et al., 2019; Alaloul, 2022).

The term "Water 4.0", also known as "Intelligent Water Networks", refers to incorporating digital automation and cutting-edge technology into water management systems to enhance the efficacy and effectiveness of water distribution (Adedeji et al., 2022). Economic efficiency, environmental protection, process optimisation, regulatory compliance, and optimisation of raw water intake, pumping energy consumption, reservoir water quality, and water purification systems are only some of the advantages that may be gained by adopting this strategy in the water business. Water 4.0 solutions can potentially boost the economic value of water utilities by reducing operational costs, increasing revenues, and improving return on investment. Water service and ecosystem optimisation improved operational security, and better human resource management are other ways in which it promotes environmental sustainability. Better efficiency, real-time monitoring, predictive capacities, the incorporation of big data, collaboration and cooperation between practitioners, stakeholders, policymakers, and academics are just some of the prospects presented by the Water 4.0 framework in water management (Grooms, 2018; Joseph et al., 2022). Water 4.0 improves service delivery, lowers costs, and produces better environmental results by utilising information science and technology to increase water management systems' efficiency, sustainability, and resilience.

However, Water 4.0 confronts several obstacles, including a lack of knowledge, high prices, risks, uncertainties, limited data availability and accessibility, fragmentation and heterogeneity of water governance and management, and a lack of cooperation and coordination across diverse players and sectors. Planning, change management, and security are essential for a smooth transition to digital water management systems and Industry 4.0 technology. More study, experimentation, and creativity are required to further the development and future course of Water 4.0. In light of the Fourth Industrial Revolution and the accompanying cybersecurity threats, future research should identify, assess, and mitigate new cybersecurity issues and possibilities in smart water management (Sedlak, 2014; Alabi et al., 2019; Adedeji & Hamam, 2020). While population growth, urban sprawl, global challenges, the transition to digital water management systems, and ensuring the security of digital systems from cyber-attacks all present opportunities for the water industry, Water 4.0 also presents challenges.

12.1.1 Policy and Regulation

Stringent regulations must be implemented to protect the privacy and integrity of the information gathered and communicated by IoT devices and smart water management systems. Protect sensitive information about water from hackers and data breaches by requiring encryption, authentication, and secure communication protocols. Furthermore, standardisation and compatibility amplify the need to create and enforce standards for IoT devices, communication protocols, and data formats to facilitate easy data sharing and use. Promote the creation of standardised, open interfaces that will provide seamless interoperability between products from various manufacturers.

In water management systems, provide explicit criteria for data ownership and exchange. Improve management and policymaking using data-driven insights by encouraging water utilities, governments, and academic institutions to enter open data-sharing agreements. Amplify sandboxes in regulation and other incentives for new ideas by building innovative centres where IoT and smart water management systems may be tested and improved without meeting strict compliance standards. Also, it offers financial or tax breaks to promote cutting-edge technology that improves the effectiveness and longevity of water distribution systems. The protection of vital infrastructure from cyber threats and assaults relies on the creation of comprehensive cybersecurity guidelines that are adapted to water management systems; it is, therefore, necessary to require regular security audits, penetration tests, and incident response strategies for water networks that use the IoT.

Managing and guiding data described in data collection, storage, processing, and retention should all be governed by established rules. Promote cutting-edge analytic tools and machine learning methods to mine data for actionable intelligence and aid in making sound judgments. Spread the word about the merits of IoT-enabled water management and the need to conserve water supplies via public awareness campaigns aimed at consumers, companies, and other stakeholders. Therefore, people should be encouraged to adopt water-saving habits and reduce their water use whenever possible. Furthermore, new technology can be quickly incorporated into existing water management systems by enacting legislation that can adapt to the ever-quickening pace of technological change. In the same way, regular policy reviews should be incorporated to keep up with the ever-changing business climate.

Encourage coordination between public entities, water companies, service providers, academics, and non-profits to create comprehensive solutions for smart water management. This can be done by forming expert panels or committees to collect feedback on developing IoT adoption and water management policy. Also, it is important to provide resources for research and development initiatives that investigate novel IoT-based approaches to water management issues by funding studies that aim to remove technological hurdles, such as those related to sensor precision and power consumption. Likewise, it helps people worldwide work together and share information on smart water management's finest methods, regulatory strategies, and cutting-edge technologies. Join other nations to solve the world's water problems using the IoT and other advanced technological methods. Instigate means of checking that legislation about the IoT and intelligent water management is being followed. Encourage people to follow the rules, and enforce punishments for those who do not.

A complete legal framework that considers security, privacy, interoperability, innovation, and stakeholder engagement is necessary to properly use the promise of IoT and intelligent water management technologies. The goal of this framework should be to protect vital water infrastructure and supplies while encouraging technological progress. With these guidelines in place, governments and businesses can be sure that IoT and associated technologies will be used in water management systems in a way that promotes efficiency, conservation, and resilience in the face of changing conditions.

12.2 CONCLUSION

In conclusion, this book on Water 4.0, which integrates cutting-edge technologies for efficient water management, provides a deep dive into the idea of Water 4.0 and its potential to transform water management. It highlights the importance of AI, automation, CPS, and the IoT in water distribution systems. Improved operational efficiency, sustainability, and dependability are benefits highlighted in this book as reasons to embrace Water 4.0. Many uses of Water 4.0 are explored, from controlling water pressure to detecting leaks in pipes to keeping track of water quality. These programmes may increase the efficiency and durability of water distribution networks. Several difficulties with implementing Water 4.0 are acknowledged in this book, such as the high price of adoption of IoT technology, cybersecurity risks, data quality and unpredictability, energy solutions, and regulatory regulations. It stresses the need for stakeholders, governments, and academics to work together to encourage Water 4.0 adoption in the face of these obstacles.

This book suggests using MPC and lossless compression techniques to guarantee efficient data management. CPS are also highlighted for their value in water quality monitoring, energy management, and leak detection in water distribution networks. Several plans for bringing Water 4.0 to fruition are outlined in this book. To encourage the widespread implementation of Water 4.0, it stresses the need for strict rules, data ownership and interchange requirements, and monetary incentives. It also proposes flexible regulations to enable the adoption of new technology into current water management systems and public awareness programmes to promote water-saving practices. Security, privacy, interoperability, innovation, and stakeholder participation are all stressed throughout this book, as is the need for a complete legal framework. Protecting critical water infrastructure and supplies, increasing efficiency, and reducing water waste are all goals that might be achieved with the help of an appropriate framework for implementing IoT and intelligent water management technology.

This book is an excellent resource for academics, government officials, and business leaders interested in Water 4.0. It sheds light on the merits, difficulties, and solutions of incorporating cutting-edge technology into water management systems. This book establishes the groundwork for future research and innovation in innovative water management by addressing significant concerns and suggesting practical solutions.

REFERENCES

Adedeji, K. B., & Hamam, Y. (2020). Cyber-physical systems for water supply network management: Basics, challenges, and roadmap. *Sustainability*, *12*(22), 9555.

Adedeji, K. B., Ponnle, A. A., Abu-Mahfouz, A. M., & Kurien, A. M. (2022). Towards digitalization of water supply systems for sustainable smart city development—Water 4.0. *Applied Sciences*, *12*(18), 9174.

Alabi, M. O., Telukdarie, A., & Janse van Rensburg, N. J. (2019). Water 4.0: An integrated business model from an industry 4.0 approach. In *Proceedings of the 2019 IEEE International Conference on Industrial Engineering and Engineering Management (IEEM)* (pp. 1576–1580). IEEE. https://doi.org/10.1109/IEEM44572.2019.8978859

Alaloul, W. S. (2022). *Cyber-Physical Systems in the Construction Sector*. CRC Press.

Chen, H., Xiong, P., Schwan, K., Gavrilovska, A., & Xu, C. (2012). A cyber-physical integrated system for application performance and energy management in data centres. In *Proceedings of the 2012 International Green Computing Conference (IGCC)* (pp. 1–9). IEEE. https://doi.org/10.1109/IGCC.2012.6322261

Grooms, K. K. (2018). *Book Review: "Water 4.0: The Past, Present, and Future of the World's Most Vital Resource"*. World Scientific.

Joseph, K., Sharma, A. K., & Van Staden, R. (2022). Development of an intelligent urban water network system. *Water, 14*(9), 1320.

Kagermann, H., Wahlster, W., & Helbig, J. (2013). Recommendations for implementing the strategic initiative Industrie 4.0. *Final Report of the Industrie, 4*(0), 82.

Kim, J. J., Taha, M. M. R., & Ross, T. J. (2014). Binary damage classification in SHM using possibility distributions. In *Vulnerability, Uncertainty, and Risk: Quantification, Mitigation, and Management* (pp. 175–183). American Society of Civil Engineers (ASCE). https://doi.org/10.1061/9780784413609.018

Lesch, V., Züfle, M., Bauer, A., Iffländer, L., Krupitzer, C., & Kounev, S. (2023). A literature review of IoT and CPS—What they are, and what they are not. *Journal of Systems and Software, 200*, 111631.

Maier, H. R., Kapelan, Z., Kasprzyk, J., Kollat, J., Matott, L. S., Cunha, M. C., Dandy, G. C., Gibbs, M. S., Keedwell, E., & Marchi, A. (2014). Evolutionary algorithms and other metaheuristics in water resources: Current status, research challenges and future directions. *Environmental Modelling & Software, 62*, 271–299.

Mallik, S., Mishra, U., & Paul, N. (2021). Groundwater suitability analysis for drinking using GIS-based fuzzy logic. *Ecological Indicators, 121*, 107179.

Pons, M., Valenzuela, E., Rodríguez, B., Nolazco-Flores, J. A., & Del-Valle-Soto, C. (2023). Utilization of 5G technologies in IoT applications: Current limitations by interference and network optimization difficulties—A review. *Sensors, 23*(8), 3876.

Rai, D., Sharma, S., & Bhuria, V. (2012). *Fuzzy Speed Controller Design of Three Phase Induction Motor*. https://www.ijetae.com/files/Volume2Issue5/IJETAE_0512_24.pdf

Scholes, E., Verheyen, V., & Brook-Carter, P. (2016). A review of practical tools for rapid monitoring of membrane bioreactors. *Water Research, 102*, 252–262.

Sedlak, D. (2014). *Water 4.0: The Past, Present, and Future of the World? S Most Vital Resource*. Yale University Press.

Tim, Y., Chiew, T. K., Lim, H. M., Teo, C. H., & Ng, C. J. (2023). Design process knowledge for crisis-driven information systems solutions: Insights on building digital resilience from an action design research study. *Information Systems Journal, 33*(6), 1343–1369. https://doi.org/10.1111/isj.12457

Wani, J. A., Sing, S., Malik, A., & Singh, P. K. (2023). An analysis study of security threats, vulnerabilities challenges of cyber-physical systems. In P. K. Singh, S. T. Wierzchoń, W. Pawłowski, A. K. Kar, & Y. Kumar (Eds.), *IoT, Big Data and AI for Improving Quality of Everyday Life: Present and Future Challenges: IOT, Big Data and AI for Improving Quality of Everyday Life: Present and Future Challenges* (Studies in Computational Intelligence, vol. 1104, pp. 153–169). Springer. https://doi.org/10.1007/978-3-031-35783-1_9

Zhang, W., Li, H., Li, Y., Liu, H., Chen, Y., & Ding, X. (2021). Application of deep learning algorithms in geotechnical engineering: A short critical review. *Artificial Intelligence Review, 54*(8), 5633–5673. https://doi.org/10.1007/s10462-021-09967-1

Zhou, J., & Wang, M. (2023). The role of government-industry-academia partnership in business incubation: Evidence from new R&D institutions in China. *Technology in Society, 72*, 102194.

Index

T

U

V

W

Z

For Product Safety Concerns and Information please contact our EU
representative GPSR@taylorandfrancis.com
Taylor & Francis Verlag GmbH, Kaufingerstraße 24, 80331 München, Germany

www.ingramcontent.com/pod-product-compliance
Ingram Content Group UK Ltd.
Pitfield, Milton Keynes, MK11 3LW, UK
UKHW020200200926
475702UK00004B/543

* 9 7 8 1 0 4 1 0 3 1 5 6 7 *